THE WAR ON WINE

CULTURAL ECOLOGIES OF FOOD IN THE TWENTY-FIRST CENTURY
Tom Hertweck (University of Nevada, Reno) and
Iker Arranz (California State University, Bakersfield), *Series Editors*

As we move deeper into the twenty-first century, people around the globe have become increasingly aware of the way their food choices produce ecologies of effects, environmentally and otherwise. *Cultural Ecologies of Food in the Twenty-First Century* invites manuscripts that, using a truly interdisciplinary framework, parse the complexities of contemporary food culture. Encompassing any characteristics of food and drink, from their agricultural or technological production to their traditional or market-based consumption, and including their systems of waste and the cultures of thought that surround them, works in the series will uncover how humanity's daily eating is constellated within and among diverse bodies of knowledge. The series encourages the work of specialists who are eager to relate their learned understanding of eating to those outside their own discipline. From the politics, economics, and scientific practices of agriculture at any scale, to the systems of promotion, distribution, and consumption that make food salable, to the representational economies of value that tell us what is good to eat and when: any transdisciplinary approach that brings food into focus will be considered.

Of particular interest are those manuscripts that include deep place-based perspectives or the environmental effects of how we eat as part of their investigations, including those that attempt either to pose well the questions food scholars and real-world eaters must face as well as to answer the extant dilemmas of our time. The series also welcomes projects that tackle the global reach of food systems and comparative studies of producing, eating, and food thought, as well as those studies that attempt to ground their work in the historical systems that inform our present moment.

Through a Vegan Studies Lens: Textual Ethics and Lived Activism
by Laura Wright

*Farm to Form: Modernist Literature and Ecologies
of Food in the British Empire*
by Jessica Martell

*The War on Wine: Prohibition, Neoprohibition,
and American Culture*
by Victor W. Geraci

THE WAR ON WINE

Prohibition, Neoprohibition,
and American Culture

VICTOR W. GERACI

UNIVERSITY OF NEVADA PRESS | *Reno & Las Vegas*

University of Nevada Press | Reno, Nevada 89557 USA
www.unpress.nevada.edu

FIRST PRINTING

Cover design by Trudi Gershinov/TG Design
Cover photograph: Chicago Daily News collection (DN-0772930); Chicago History
Museum

LIBRARY OF CONGRESS CATALOGING-IN-PUBLICATION DATA
Names: Geraci, Victor W. (Victor William), 1948– author.
Title: The war on wine : prohibition, neoprohibition, and American culture / Victor W.
  Geraci.
Other titles: Cultural ecologies of food in the 21st century.
Description: Reno : University of Nevada Press, [2023] | Series: Cultural ecologies of
  food | Includes bibliographical references and index. |
Summary: "*The War on Wine* explores the development of an American wine culture
  from the colonial period up to the present, with close consideration of efforts by
  prohibitionist interests and others to prevent it. Throughout American history the
  prohibition and restriction of alcohol, including wine, has been part of what we now
  call culture wars. After losing the Prohibition Constitutional Amendment, anti-alcohol
  forces rebranded themselves as neoprohibitionists dedicated to the restriction of
  alcohol usage and they touted themselves as the counter-voice to alcohol organizations
  like the Wine Institute led by John A. De Luca from 1976 to 2013."—Provided by
  publisher.
Identifiers: LCCN 2022062259 | ISBN 9781647791148 (paperback) | ISBN
  9781647791155 (ebook)
Subjects: LCSH: Wine and wine making—United States—History. | Wine industry—
  United States—History. | Wine and wine making—Law and legislation—United States.
  | Alcohol industry—United States—History. | Alcohol—Law and legislation—United
  States. | Prohibition—United States—History.
Classification: LCC TP557 .G48 2023 | DDC 663/.200973—dc23/eng/20230302
LC record available at https://lccn.loc.gov/2022062259

*To my grandfather Michelangelo, my father,
Victor, and Uncles Joe, Peter, George, and
Michael, who transferred their Sicilian food and
wine heritage to their American wine culture*

Contents

Acknowledgments

Between 2003 and 2013 I served as the food and wine specialist and associate director for the University of California (UC) Berkeley, Oral History Center. It was during this time that I had the opportunity to meet and complete extensive recorded and transcribed interviews with John A. De Luca, Wine Institute president and chief executive officer (CEO), as he finished his career at the helm of the influential wine organization.[1] During our interviews, our personal ties to Sicily, food, and wine heritage issues enabled us to build both a personal and a professional bond. I did not always agree with all of his historical assessments with regard to the role of wine and alcohol in American history, but I quickly realized his key role in brokering a more centrist understanding of the role of language in any discussion around wine, health, politics, and culture.

De Luca brought a deep set of cultural food and wine traditions to his Wine Institute job. His immigrant Sicilian parents had taught him their food and wine ways, and his 1930s New York City Lower East Side "Little Italy" neighborhood cemented them into his daily life. This food heritage deepened as he studied and worked in Rome. For his entire life he consumed a fresh, local, and seasonal Mediterranean diet where wine was an integral beverage served with meals. Not surprisingly, by the end of the twentieth century many Americans had embraced both Italian foodways and French *terroir* food traditions and brought them into the American canon of dietary practice.

During the years of researching and writing this book, I have been lucky enough to spend months in both northern and southern Italy to view and experience the modern versions of Italian slow food and wine traditions. In vineyards, wineries, olive groves, dairies, cheese plants, farmers markets, cooking schools, and pig farms and processing facilities, I began to understand these century-old food and wine traditions. To ensure quality and traditional growing and production methods, European nations have created the Protected Designation of Origin, or Denominazione di Origine Protetta in Italy. These strict rules and traditions protect foods and wines whose specific characteristics depend upon a regulated interaction between natural geographical factors and human intervention that together produce a site-specific product. Producers must follow strict rules, and compliance is enforced by independent consortia. Daily cultural traditions then encourage and enhance guidelines for use of the products. Deeply embedded in these foodways is the idea of wine as a part of meals.[2] De Luca understood and took pride in the national foodways of his familial homeland.

But it would be his ability to politically negotiate a common ground between the deeply held beliefs of pro-wine advocates and anti-alcohol forces that sealed his place in the history of American wine. For more than three decades, he directly influenced the continuance of the notion of an American wine culture built on the ideals of European wine cultures that promoted moderation and wine as part of a healthy diet. During his leadership years, in an effort to protect his food and wine traditions and an American wine industry, he directly battled anti-alcohol forces in the American alcohol culture war.

Due to the thrust of the book's main argument about how an industry addressed an American approach to alcohol, wine in particular, I do not attempt an in-depth review of the contemporary historiography of race, class, and gender. While these issues are of timely concern, I am not abandoning or downplaying their importance in academic discussions. However, although these arguments are important for theory-based historians, they do not necessarily enlighten this story of creating an American alcohol policy forged and mitigated by the tenants of capitalism, progressivism, politics, and religion. The aim of the story is to illuminate, for the general public, a chronicle of the nation's unease with alcohol issues and policies that have vacillated between condemnation and an uneasy acceptance of consumption. A discussion of race, class, and gender would force the general topic into a rabbit hole of complex discussions that would cloud the everyday understanding of the components of the manuscript. I have attempted to generalize these complex theories out of a concern that they would distract from the book's main points. As a traditionally trained public historian, I tell a story of change over time, for as large an audience as possible, and attempt to provide access to those who would be interested in the battle for an American wine culture.

As De Luca and I concluded his interviews, he issued a warning to the industry and wine drinkers that they needed to stand vigilant in what he saw as the never-ending anti-alcohol battle that sought to end or restrict the drinking of alcohol, wine in particular. De Luca spent his wine-industry career navigating a path for an American wine culture in a nation with many determined anti-alcohol enthusiasts. For three decades he fought numerous political, science, health, and education battles and carried the banner for the continuance of an American wine culture. Over the past few years, as I continued my food and wine research and writing, I ruminated about his warning, and the end result is this book.

Researching and writing during a pandemic slowed the process but did not keep me from telling this story. As in all research projects, many people and organizations assisted me in setting down my account of events. First would be a thanks to Martin Meeker, past director of the UC, Berkeley, Oral History Center, for providing internet access to their recorded and transcribed food and wine interviews. De Luca's interview provided a starting point for my more than one hundred hours of personal recorded conversations that John and I, prepandemic, completed between February and July 2019. The additional interviews centered on the struggle for an American wine culture based upon European food traditions that immigrants had brought to America.

Thank you also to the following people for assisting in the writing of this book: De Luca served on the Advisory Board for Cañada College, and as a courtesy past president Jamillah Moore, EdD, provided us with a spacious and quiet conference room to spread out research materials and conduct our interviews. This proved to be an invaluable luxury and greatly assisted me in pulling the story together. Regretfully, the pandemic year of 2020 moved us from in-person to telephone and electronic information sharing. A special thanks goes to University of Nevada Press staff members, the late Margaret Dalrymple for her unwavering belief that this book had an important story to tell, and Virginia Fontana for her amazing skills shepherding the book to completion. Special thanks also extended to Kathleen Chapman whose copyedits improved the final story.

The biggest thank you goes to the De Luca family, who assisted in all aspects of the project. John and his wife, Josephine, brought me into their home, fed me meals, and allowed me to spread out family photographs and documents in their kitchen and living room. In these meetings we discussed the materials, and they provided me with invaluable feedback as I stitched the narrative together. A special thanks to John's daughter, Carla, for assisting as a go-between for many of my questions, and thanks to publisher, editor, and past CNN CEO Tom Johnson, who read and commented on an earlier draft.

Last, but never least, a thanks to my family and friends for listening and responding to a never-ending barrage of Sicily and wine stories, many about material in this book. As a reward I took my kids, grandkids, and their spouses to Sicily to visit our family village of Piana degli Albanesi in the province of Palermo. I also offer a special remembrance to two scholars, Otis L. Graham and Gordon Bakken, who mentored me in the ways of the historian. I am sure they are bugging God for funding to complete some heavenly book project.

—VICTOR W. GERACI, Murrieta, California

Introduction

L'acqua fa male, ma'o vino fà cantà.
—Italian proverb, "Water may make you ill, but wine makes you sing."

As I sat at my computer writing this book, I looked forward to the fact that later I would have a glass of wine for dinner or wine with friends at a bar, restaurant, or winery. Growing up in a Sicilian American grape-sharecropping family, I became accustomed to having wine at the dinner table or as a vital part of family social gatherings. Being a Californian I can purchase my favorite wines from numerous states or countries at convenience stores, grocery stores, liquor stores, the internet, or direct from the winery, pretty much any day or time that I want. But this is not true for many Americans, and just as important are the many people who believe, for religious or personal reasons, that I should have restricted access to my favorite mealtime beverage. There are those who feel that any consumption of alcohol should be illegal or at least regulated and heavily taxed so as to address alcoholism by reducing consumption. These feelings toward alcohol are not new and to this day directly affect the way Americans purchase and drink alcohol. The end result has been a national on-again, off-again relationship with spirits and how to handle excessive drinking. Unlike European countries like France, Italy, Spain, and Germany, where wine is a celebrated national drink, Americans struggle with allowing free access to wine because of its image as a devil's brew that can effect countless societal problems.

The history of wine in America is a tale of capitalist production, consumer experience, and anti-alcohol forces seeking prohibition or restrictions on all alcohol. For Americans the idea of a wine culture built on moderation and wine as part of a healthy diet has been part of a national culture war. For most of the nation's history, with the exception of thirteen years of federal prohibition, the issue of alcohol has been left to individual states. One of the most common arguments made by those wanting cultural issues decided at the state level is that regional governments should pass their own laws and make them restrictive or nonrestrictive in accordance with the desires of their electorate. But history has taught Americans that issues of individual rights bearing on social morality cannot be contained within state boundaries. States' rights have always resulted in demands for federal action to guarantee the laws and desires of individual states. After the 2016 presidential election and the establishment of a conservative Supreme Court, Americans again faced calls for state and federal control of cultural issues dealing with LGBTQ rights, same-sex marriages, gun rights, abortion, birth control, educational curricula, the environment, and religion. The result has been political culture wars where some states promote local legislation on social issues, while other states promote federal policies and laws. Alcohol, now and in the past, has always been part of these culture-war struggles.

The nation has a precedent for how to deal with hot-topic cultural ideals that were turned over to the individual states. One only needs to look at the American struggle with slavery and the very real constitutional challenge of a myriad of competing state laws. Even as individual states chose different paths on slavery, the question of how to manage relations between slave and free states and travel across state lines became a persistent problem. The resulting 1850 Fugitive Slave Act was a attempted legal solution to the moral issue of slavery. The act allowed California to enter the Union as a free state, but required all states, even those that did not allow slavery within their boundaries, to cooperate with the forcible return of escaped enslaved people. As the struggle over slavery increased, compromise forced antislave states to deny human rights to enslaved people even if they lived in free states. Allowing slavery anywhere required protecting slavery everywhere. When South Carolina seceded from the Union in 1860, one of its grievances was that the federal government had not enforced the Fugitive Slave Act.

Could this same approach apply to current culture wars that divide the nation on issues such as abortion, LGBTQ rights, immigration, guns, marijuana, education, and alcohol? Before Prohibition each state decided its own alcohol policies and laws, and it is only with the Eighteenth Amendment (Prohibition) in 1919 that the nation had a federal alcohol policy. When Repeal returned alcohol policies back to the states, the wine industry faced forty-eight (later fifty) sets of state laws that varied between acceptance and outright prohibition. Today some Americans still consider wine to be an illicit drug. The battle for control of alcohol policies and laws has thrown wine into an active alcohol culture war.

To fully appreciate the complete story of the struggle to build an American wine culture, it is important to understand America's political conundrum with alcohol. The subject has a deeply layered historical context beginning in the post–Civil War years when southern European immigrants to the United States brought strange languages, Catholicism, food, and wine cultural traditions to their new home. These traditions and heritage often offended some Americans, but determined individuals fought to retain their customs for themselves and future generations of their family. Historian Lisa McGirr, in the introduction to her book *The War on Alcohol: Prohibition and the Rise of the American State,* addresses the heart of the issue when she states: "The war on alcohol was a prime example of a recurring theme of United States mass politics. The nation's powerful traditions of evangelical Protestantism and its freewheeling brand of expansive capitalism emerged in tandem—and in tension with one another. This combination of forces periodically fueled moral crusades among men and women unsettled by social conflict and change. These reformers turned to the state to stabilize the social order, and secure their place within it with strong doses of coercive moral absolutes."[1] In reality these moral anti-alcohol crusaders viewed themselves as saviors of America's image as a beacon on the hill and battled the chaos of their rapidly changing nation brought on by a massive influx of immigrants and their traditional European beer- and wine-drinking cultures.

Clouding the issue was the fact that wine is one of the oldest commercially traded commodities in world history. Throughout history grape growers and winemakers have taken laborious measures to continually supply consumers who viewed wine as an everyday part of their pharmacopoeia, religious ceremonies, celebrations, and mealtime needs. As a result, Europeans developed wine cultures whereby most people consumed wine daily for their meals and social occasions, and by the Middle Ages they had created elaborate vineyard and winery businesses to supply their needs. During the Age of Exploration European explorers then brought their wine culture to North America, where for more than two centuries they failed to develop a viable commercial wine industry that could accommodate an American wine culture.

As America transitioned from colonial governance to becoming a republic, untold attempts to create a wine culture failed in the North, South, and Midwest due to climate, pests, diseases, wars, and depressions. But strong cultural wine traditions and business opportunities pushed vintners to not accept defeat and to seek alternatives to make wines better, more plentiful, and profitable. As their wine-making learning curve progressed, they learned to match specific grapes to specific regions and created more efficient and profitable ways to produce, store, market, and distribute wines. But they failed to create a viable commercial wine industry capable of supporting an American wine culture. Despite their failures, many determined oenophiles continued their quest into the middle of the nineteenth century and drew upon the ideals of capitalism and Manifest Destiny to spark a new excitement for developing an American wine industry. Many believed that the new West Coast territories along the Pacific Ocean provided the answer with their historic precedents and grape-friendly climate and geography.

As the American quest for wine reached the new state of California, entrepreneurs quickly took control of the industry and ushered in a new wine-business era built on the tenets of Gilded Age agribusiness. By 1900 this new California wine industry prospered, Americans developed their first wine culture, and the wines produced gained both national and global recognition. But just as their hopes for a commercial wine industry prospered, the nation reversed course, and in 1919 evangelical believers, anti-immigrant forces, and anti-alcohol enthusiasts successfully sponsored the Prohibition Amendment to the Constitution. Over a thirteen-year period, Prohibition destroyed the commercial wine industry, and the nation lost its first wine culture.

After the 1933 Repeal Amendment overturned the Prohibition Amendment, the cultural passion for wine and its success as a trade item increased, and a tenacious generation of entrepreneurs plotted a new course for American wine. They began the process of rebuilding the industry, but in a short time they realized that their advertising, educational, and modernized production processes could not keep pace with increasing domestic demand. Over the next three decades Italian American wine-making families, corporate alcohol producers, and the newly established Wine Institute came together to meet increased demand for wine by modernizing and expanding the industry.

Their successes in rebuilding the modern wine industry and American wine culture did not go unnoticed by anti-alcohol proponents, who were still angry over the repeal of Prohibition. In response, anti-alcohol forces spent the last half of the twentieth century intensifying their efforts to find new ways to restrict, regulate, and tax alcohol. They may have lost the prohibition battle, but they swore to never give up the fight against alcohol and their belief that it was an illicit drug.

As the industry came into its own over the latter half of the twentieth century, wineries became a vital part of the new post–World War II military-industrial economy. The only downside to the California wine success was the fact that over these five decades, growers and vintners struggled to keep pace with expanding consumer demands. At first they utilized the tried-and-true historic solution of simply planting more vineyards. This proved to be problematic, as cheap vineyard land disappeared, forcing a new cadre of wine businesspeople to seek state-of-the-art solutions. To boost production they looked to new regions and consolidated and modernized the industry. Vintners shifted from traditional trial-and-error systems to a wine-by-design model whereby wine businesspersons utilized agribusiness techniques dependent upon science, technology, and marketing to achieve both vineyard and winery production goals while maximizing profits.[2] Overall, this California adaptation of the capitalist tenets of agribusiness produced a modern wine industry that first dominated the American market and then became an integral part of the global wine industry. America had established its second wine culture.

Winemakers, for good or bad, had developed a vintibusiness style that ruled the day.[3] Both grape growers and winemakers now spoke of things like cool-weather vineyards, temperature-controlled fermentation, irrigation, chemical pest and disease control, trellis and canopy design, leaf pulling, grape thinning, new clones, and machines for picking, spraying, and pruning. University-trained winemakers and vineyard managers employed new scientific discoveries and quickly learned how to act like megacorporations designed to bring efficiency and consistent quality to wine production. The new American wine culture now had an infrastructure capable of supplying their needs. Success, on the other hand, fostered increased animosity between capitalists and religious anti-alcohol supporters. As the industry expanded, tensions became more strained as anti-alcohol forces vocally condemned corporate winemakers as evil purveyors of booze. In response, they sought new ways to diminish corporate profits and control the use of alcohol.

While northern Italian immigrants played a key role in rebuilding the wine industry after Prohibition, it was the first-generation Sicilian American John A. De Luca who drew upon his training and Italian heritage to ensure that wine continued to have a place at the American table. De Luca brought the idea of an American wine culture to the mainstream social and political discussion by emphasizing his heritage and cultural traditions as an Italian American and openly courted politics and science as avenues for verifying his beliefs. His use of wine's history, culture, and science permeated all aspects of his life and to a great extent influenced his leadership role at the Wine Institute.

As a result, between 1975 and 2013 De Luca served as a national spokesperson and carried the banner for defending America's wine culture against the continuous attacks of anti-alcohol forces. History had taught him that America was locked in a battle between neoprohibition and the idea of an American wine culture. More important, he reminded Americans that the battle between pro- and anti-alcohol forces had a long history and was far from over. He challenged both producers and drinkers to engage with anti-alcohol forces and continue the struggle for continuance of an American wine culture.

THE WAR ON WINE

The Early Republic's Failure to Establish a Wine Culture

For the present I confine myself to the physical want of some good Montepulciano. . . . [T]he warm season will be so fast advancing, when you receive this, that no time will be to be lost. Perhaps I may trouble you annually to about the same amount, this being a very favorite wine, and habit having rendered the light and high flavored wines a necessary of life with me.

—Letter from Thomas Jefferson to Thomas Appleton
(American consul in Livorno, Italy), January 14, 1816

The history of wine is a tale of capitalist production and consumer experience, and the story exemplifies ancient and modern human needs to make life better in an increasingly complex world. Throughout the past eight thousand years of Western history, wine embarked on a journey beginning in northern Iran and then meandered in a westward direction to embrace Egyptian, Greek, Roman, European, and American wine enthusiasts.[1] Integral to the story has been the business of viticulture, or cultivation of grapevines, and viniculture, the study of wine making.

Throughout this journey oenophiles studied and praised wine for its role as a preserved food and as a beverage for feasting and celebrating. Humankind quickly developed a wine culture built upon the concepts that wine provided them with one of their first domesticated crops, a cultural food heritage, a medicinal substance, a major trade item, and a symbol of their God. Thus, wine culture has played itself out in the political and economic dramas of the Greeks, Romans, early Christians, European nation-states, the New World colonies, modern nation-states, and present-day America. Because of this, wine became historically intertwined with the politics, legal frameworks, culture, and prevailing economic systems throughout the world. By the fourteenth century, viticulture served as one of many symbols of power for northern and western European ruling classes. More important, by the seventeenth and eighteenth centuries, wine businesspersons experimented with the mercantile (and later capitalist) ideas of capital, credit, new technology, and marketing.[2]

Historical geographer Tim Unwin describes the cultural landscape of viticulture as "an expression of transformations and interactions in the economic, social, political, and ideological structures of a particular people at a specific place."[3] In America, this history of wine manifested itself as an intertwined set of interests, those of a wealthy class in need of an artistic hobby; of a wine industry seeking profits; of a consumer-focused middle class seeking symbols of the good life; of admirers of the Jeffersonian agrarian myth; of temperance and religious groups wishing to legislate morality around alcohol; and of a nation in search of a national public health policy.

WINE EMIGRATES TO ENGLISH COLONIAL AMERICA

Viticulture arrived in the Americas during the Age of Discovery (1500–1750) as European explorers and immigrants brought their wine-drinking traditions to the New World. They found that none of the Native American peoples had any knowledge of wine, or for that matter, of fermented spirits. Over the next three centuries, these western immigrants utilized native grapes and later introduced European vines (*Vitis vinifera*) in countless attempts to develop a successful American wine industry capable of supporting a local wine culture.

These early European explorers and settlers documented the possibilities for and attempts at grape growing and wine making. In 1524 Giovanni da Verrazano, while off the coast of North Carolina, found "many vines growing naturally, which growing up, took hold of the trees as they do in Lombardy, which if by husbandmen they were dressed in good order, without all doubt they would yield excellent wines." A decade later, Frenchman Jacques Cartier explored the St. Lawrence Gulf and described one island as having "many good vines, a thing not before of us seen in those countries, and therefore we named it Bacchus Land." These early discoveries led to the first wine making in North America in 1564 when French Huguenots fermented local scuppernong grapes in Florida. They failed to grow enough food for survival, but were able to produce twenty hogsheads of wine.[4]

Viticulture arrived in English North America in 1607 when

Jamestown settlers planted grapevines, and by 1609 samples of their Virginia wine reached Europe.[5] The New World wine pleased King James I, who hated tobacco, and he subsequently urged settlers to plant vineyards. Following the king's desires, in 1619 the Virginia Company enacted a law that required each householder to plant and maintain ten vines yearly until they had a vineyard. To assist in the endeavor the company provided French vignerons to instruct and oversee the vineyards and wine making. Despite valiant efforts, they fermented little wine because their native vineyards produced poor-quality grapes. Two more attempts in the 1630s and 1650s failed, and by midcentury, dreams of a get-rich-quick wine industry forced wealthy English colonists to continue to import wine from Italy, Spain, and France. Dreams of an American wine culture stalled.

When England adopted an official policy that no longer supported the development of a colonial wine industry, entrepreneurs decided to move their wine-making attempts to the agricultural southern colonies.[6] To garner support for their cause, they approached the South Carolina House of Commons. In response the Commons offered a 100-pound prize to "the first person who shall make the first pipe of good, strong-bodied merchantable wine of the growth and culture of his own plantation."[7] In 1758 Robert Thorpe, Esq., laid claim to the prize and incentivized a new flurry of experimentation in grape growing and wine production.

Regretfully, plans to build the American wine culture languished as vineyards in the southern colonies proved deficient because of the selection of inappropriate varieties for climatic conditions, pests, disease, political unrest, and, later the American Revolution. In 1766 Dr. Andrew Turnbull, ex-British consul to Smyrna, imported Greek and Mediterranean cuttings and hired immigrant viticulturalists to plant his Florida vineyards. His attempt failed as his Mediterranean varieties succumbed to tropical diseases. North Carolina growers utilized native grapes for a European-style wine industry that failed as their vineyards contracted numerous diseases. Louisiana settler Colonel Ball produced enough wine in 1775 to send a sample of his claret to King George III. The project faltered as an uprising by local indigenous people put an end to his enterprise. At the same time, Virginia offered a prize for planting wine-grape vineyards and had no takers.

Thomas Jefferson made one final prerevolutionary attempt at

viticulture in the southern colonies when he contracted Philip Mazzei, a Florentine physician and merchant, to plant vineyards of European varietals near his Monticello home in Charlottesville, Virginia. The Mazzei project ended when the American Revolution turned Jefferson's attention away from his agricultural endeavors. Yet colonial viticultural failures in Virginia, Georgia, Louisiana, the Carolinas, and Florida did not discourage Americans from attempting to build a wine industry.[8]

Undaunted wine entrepreneurs now turned their attention to the middle and northern colonies, only to again suffer setbacks as growers tried to establish European wine-grape vineyards. Colonel Benjamin Tasker Jr. and Charles Carroll planted experimental vineyards in Prince George County, Maryland, and by 1796 both projects ceased to exist.[9] North of the Mason-Dixon line, Pennsylvania Germans experimented with new wine and grape techniques, and Benjamin Franklin encouraged the development of native wines and offered directions for wine making in his *Poor Richards Almanac*. Throughout the 1760s and 1770s, numerous small attempts failed to produce American wines, and some declared that *Vitis vinifera* was a lost cause in America. Early attempts for a commercial wine industry and American wine culture had failed.

LIMITED SUCCESS IN DEVELOPING
THE FIRST AMERICAN WINE CULTURE

Vineyard failures in New England and the South worried those seeking to establish an American industry. Undaunted, many now looked westward to lands and climates more suited to grape growing as a possibility for a commercial industry. They were encouraged by wine-culture enthusiast Thomas Jefferson, who claimed that Americans could "make as great a variety of wines as are made in Europe," just "not exactly the same kinds, but doubtless as good."[10] This faith in the ability of America to produce wine propelled the continued quest for American wine.[11]

Frenchman Peter Legaux helped reinvigorate the American viticultural quest in 1793 by planting a grapevine nursery and 206-acre vineyard in Philadelphia. Regretfully, even with presidential support and state funding, he could not overcome the forces of nature, and by 1822 frost, disease, drought, and a caterpillar plague ended his viticultural project.

Still, a national hope for American wine continued, as growers in Maine, New York, New Jersey, Maryland, Virginia, Ohio, and Kentucky utilized Legeaux's native hybrid varieties to establish new vineyards. Some of these vines found their way to Englishman Benjamin Vaughan, in Hallowell, Maine. Vaughan left viticulture records that would later be used by Dr. James Mease, a Philadelphia physician, when he wrote about American vines in his *Domestic Encyclopedia.* As wine interest increased, prospective vintners turned to new publications on viticulture like the 1795 *A Short and Practical Treatise on the Culture of the Winegrapes in the United States of America, Adopted to Those States Situated to the Southward of 41 Degrees of North Latitude.*[12]

Trial and error began to produce an accumulated knowledge that would lead to eventual limited successes in the territories west of the Appalachian Mountains. The first of these came in 1799 when Jean-Jacques Dufour, Swiss viticulturist, started the Kentucky Vine Company. Like earlier attempts farther east, the Dufour project failed as a result of plant disease, low yields, and general bad luck. Dufour was undaunted, and in 1802 he convinced Congress to grant him land on the Ohio River where he planted vineyards as part of his New Switzerland Colony. In a very short time, Kentucky wines began to reach markets all over the United States, and growers read Dufour's book, *The American Vine-Dresser's Guide.* By the 1820s, like his predecessors', all his vineyards in Kentucky succumbed to disease.[13]

None of these failures undermined the entrepreneurial urge to establish an American wine industry. From the midst of all these delays came Major John Adlum (1759–1836), whom many would later call the "Father of American Viticulture." The Pennsylvania-born veteran of the War of 1812 settled on a Maryland farm where he watched his first vineyard of European vines succumb to insects and disease. Adlum then moved to Georgetown and initiated a correspondence with Thomas Jefferson, who suggested planting a new vineyard on native rootstock. Jefferson's belief that "wine is a daily necessity," and his detailed notes on his failures, provided Adlum with new data to plant a vineyard. Armed with scientific facts, Adlum started anew. In a short time, he had success with native *Catawba* vines grafted to European varietals. Wine devotees now labeled him the guru and chief wine propagandist of American viticulture, and many prospective growers studied Adlum's 1826–27 *The Vigneron* and *Adlum on Wine Making.*[14]

Accumulated information from numerous failures convinced die-hard wine entrepreneurs to again look southward. Numerous North and South Carolina planters sought viticulture as an alternative to diversify their tobacco and cotton slave-based agricultural system. Their initial attempts to establish vineyards failed after the South Carolina Society for Promoting Agriculture ran out of funds before finding the right combination of vines, soil, labor, and knowledge. To help establish vineyards, Nicholas Herbemont, a southern grower, urged the South Carolina Legislature in 1827 to subsidize emigration of vignerons from France, Italy, Germany, and Switzerland. Based on past failures, the legislature found that the scheme was too risky and rejected Herbemont's proposal. A similar 1828 proposal to the Georgia Legislature also failed to gain traction.

LIMITED COMMERCIAL WINE SUCCESSES— OHIO, NEW YORK, AND MISSOURI

As the list of unsuccessful viticultural projects grew, there were a few successes that utilized native or hybrid grape varietals. This kept wine enthusiasts engaged and served as the catalyst for a new round of attempts to build an industry. First in this category would be lawyer, horticulturist, and wealthy Ohio developer Nicholas Longworth. He utilized much of his wealth in 1850 to establish a modern, commercial Ohio wine industry. Longworth had moved west after his New Jersey estate had been laid to ruins by the British in the Revolutionary War. He invested heavily in Ohio lands and used his real-estate profits to "democratize wine appreciation—to bring it within the ken of ordinary citizens by treating it as something refined and gracious and as part and parcel of America's agrarian ideal."[15]

Longworth decided that using native American grapes to produce a dry, white table wine would be the best commercial way to provide American wine for the temperance movement's moderation beliefs. By using advanced marketing tactics, he also pushed against American wine snobbery, that favored European wines, and created labels that downplayed his use of native grapes and advertised his thirty-five-thousand-gallon annual production. Within a few years, the vineyard and wine business became a cutting-edge enterprise, and his wines gained national and international recognition. Longworth's successful enterprise ended, as the Civil War interrupted the industry and he failed to secure a stable supply of quality grapes.[16]

Another semi-successful, commercial wine-grape–growing venture emerged in the 1850s with the New York Finger Lakes Pleasant Valley Wine Company. In a short time it failed. Other growers continued the quest, like Frenchman Alphonse Loubat, who planted forty acres on the Brooklyn waterfront, and William Robert Prince, Long Island nurseryman, who developed a viticultural nursery and published a 1830 grape guide called *A Treatise on the Vine*.

Hopes for an American wine culture then moved farther west to Missouri. In 1859 St. Louis wine growers formed the American Wine Company. This enterprise was quickly followed by Michael Poeschel's Stone Hill Winery in 1861 that became one of the largest wineries outside of California (it operated until Prohibition and then reopened in 1965). Successes like these prompted American leaders to continue to promote an American wine culture. In 1851 Illinois senator Stephen A. Douglas enthusiastically stated that the "United States will, in a very short time, produce good wine, so cheap, and in such abundance, as to render it a common and daily beverage."[17] Georgetown's John Adlum declared that by 1900, Virginia and the Southwest would have as many acres of vineyards as France and would produce wines equal to the French. At best, the idea of an American wine culture was off to a painstakingly slow start.

America, through trial and error, still had not found the proper grape varietals and the perfect climatic region that would reduce vineyard problems. This need for a wine-grape climate did not escape those wishing to build an industry. In an 1870 *Harper's Magazine* story, writer and Ohio legislator William Joseph Flagg correctly identified America's wine problem. In the article he proclaimed, "The question of wine drinking in America revolves itself into the question of grape growing in America."[18] Enthusiasts sought new grape-growing regions that would benefit from the accumulated knowledge left behind from past failures and limited successes.

New hopes rose as vintners began moving beyond trial and error and learned to disseminate information, fund research, form organizations, vigorously market their product, lure investors, and seek governmental support. Universities jumped in and set up experimental stations, and academic departments devoted to viticulture and viniculture and exhibits at fairs and exhibitions demonstrated the newest in technology and plant material. The federal government encouraged grape growing, and by 1857 the Patent Office made a systematic effort to collect and study native vines capable of supporting a wine industry. In 1862 the newly established United States Department of Agriculture (USDA) committed resources to assist in the development of an American wine industry.

The process of building an American wine industry slowed in the 1860s as the Civil War consumed the nation and funneled resources away from grape and wine projects. This did not discourage wine enthusiasts, who realized that wars have always had a tendency to interrupt the routines and business of daily life. French vineyard peasants, after centuries of war, would have advised American wine drinkers to take solace in their belief that the "Good Lord sends a poor wine crop when war starts and a fine, festive one to mark its end."[19] For the United States, the Civil War, and its resulting Reconstruction, slowed the development of the eastern and southern wine industry. Yet there could be no doubt that the country desperately sought a stable commercial wine industry capable of supporting an American wine culture.

A NATION IN SEARCH OF AN ALCOHOL TRADITION

Failures in early American viticulture led most wine-drinking settlers to substitute new libations while they continued the search for a means to produce their own wine. Yeoman farmers followed their European traditions and developed a subsistence-plus mentality that left grape growing and wine production to gentleman farmers. This helped create an American wine mystique of a drink for a gentler class. On the other hand, President Jefferson, while representing this American genteel class, hoped to cultivate the French, Italian, Spanish, and English tradition of a wine culture for the common people. These hopes diminished as viticultural failures grew and postrevolutionary Americans turned to cheaper rum that quickly accounted for one-fifth of the value of all imports from England. Jefferson worried that distilled spirits would diminish the vitality of the nation's health and continued to push for an American wine industry. In the interim, most Americans turned to stronger drink, and the idea of wine as a moderating force seemed like a far-off dream.

Eighteenth-century Americans continued their European beverage traditions by substituting distilled spirits for wine for their meals and social entertainment. Alcohol did not present a problem for the general population. Even the Puritans had viewed alcohol as the "Good Creature of God," a holy substance, when used cautiously. According to the Rev. James Alexander of Princeton, New Jersey, "The Bible speaks well of wine, even as an exhilarate."[20] As early as 1708, religious leaders like Cotton Mather had preached moderation. He affirmed the preaching of his father—Reverend Increase Mather—on the value of moderation of this "Creature of God." Religious leaders preached the merits of alcohol in moderation and feared that drunkenness, from cheap distilled strong drinks, emanated from Satan.

Just before the American Revolution, all ages and social groups consumed alcohol, and foreign travelers reported that America was a nation addicted to drinking. The national consumption rate had reached three and one-half gallons per person (more than the present rate of consumption) per year. Nine million women and children drank twelve million gallons of distilled spirits, and three million men drank more than sixty million gallons.[21] Religious people now began to reevaluate their stance on alcohol as America became an alcoholic republic.[22] The nation had developed an alcohol problem.

As early as 1772, Benjamin Rush, a physician, concluded that distilled liquor destroyed the body's natural balance. To prove his point, Rush quoted reports from the American Continental Army's surgeon general that tied ill health of soldiers to issuance of rum. Rush believed that the military should substitute wine for distilled spirits. He reaffirmed wine as a natural and healthy food by comparing sickly American soldiers to Roman soldiers known for their vigor and vitality.[23]

To meet the new demand, American distilled-alcohol production increased, and cheap whiskey from excess domestic grain, caused by embargoes from the American Revolution and the War of 1812, helped solidify a new American drinking habit. Whiskey flourished as a "liquid asset" when grain-glutted markets took advantage of improved crop-distribution methods, and new large-scale distilleries capitalized on the fact that America suffered from impure water supplies, lack of milk, wine shortages, and expensive tea.[24]

Alcoholic beverages played a central role in the developing American lifestyle. Southern slave owners provided watered-down alcohol as a work incentive for enslaved people, parents taught moderate drinking habits by giving their children spirits, and women drank at home or in mixed company. Indigenous peoples received whiskey for payments, young men proved their manhood by overindulging, the military issued distilled spirits to sailors and soldiers, and college students relaxed in pubs. By 1830 Americans over the age of fifteen consumed more than seven gallons a year, or over two times the modern rate of consumption.[25] The American diet of fried greasy foods, butter, and eggs was consumed with whiskey.

Industrialization, modernization, rural isolation, and impersonal factory systems became ways of life and helped solidify the idea of America as a nation of drunkards. Despite Americans' increased consumption, alcohol was not considered a social problem, and informal social groups strictly enforced the regulation of antisocial drinking behavior in the tavern and embraced alcohol as a benign and healthful beverage.[26] Most Americans, like their European ancestors, viewed alcohol as a food and medication for colds, fevers, frostbite, and depression, and as a means of tension reduction, as well as a way for all social classes to enjoy frivolous camaraderie. Local governments monitored drinking through the issuance of tavern licenses and issued laws aimed at controlling public drunkenness and its related crimes of thievery, lechery, and brutality.

This national drinking binge worried many because of alcohol's adverse medical and social effects. The new concerns strengthened the position of the temperance movement, and by the 1820s concerned citizens promoted reform measures aimed at moderation. These early reformers started a national debate on the health and social values of fermented and distilled alcoholic beverages. Most people still believed in supporting policies and laws that actively encouraged drinking wine and distilled spirits in moderation. They were countered by extreme reformers, focused on the evils of alcohol, who believed in teetotalism.[27]

Most anti-alcohol groups of this period did not wish to prohibit the consumption of spirits because they realized that drinking habits had become associated with personal freedom and a modern lifestyle. In an attempt to regulate alcohol consumption, factory owners, middle-class urban Americans fighting crime, evangelical leaders preaching social responsibility, women's temperance groups hoping to save families, and political leaders collaborated to address the problem. The reformers blamed economic decay, poverty, and social upheavals on hard drink and pushed for governmental laws and policies that emphasized moderation. Few believed that prohibiting alcohol would work. They remembered futile attempts to deal with alcohol issues that faltered when officials taxed distilled spirits and faced a Whiskey Rebellion led by angry farmers accustomed to producing a grain surplus for whiskey.[28]

Most reformers favored abandoning distilled spirits and promoted the use of wine, in moderation, as a beverage for consumption with meals. The problem centered on the fact that there was no real wine industry available to supply consumer needs and that only wealthy citizens could afford higher-priced imported European wines. With a lack of domestic vineyards and wineries, the price of wine had skyrocketed to one dollar per gallon, or four times the price of whiskey. At the national level, many worried about the economic impact caused by the exorbitant prices of foreign wine that had helped worsen the American balance of payments with its European trade partners.

Despite the fact that rank-and-file Americans drank rum and whiskey, many retained the vision and dream of an American wine industry. Hopeful journalist Hezekiah Niles prophesied that in time the United States would produce all its own wine, and the situation would resolve itself.[29] The developing moderation approach to alcohol abuse brought a new commitment to establish an American wine industry for production of cheaper wine for all Americans. The fact remained that in order to have a wine culture, America needed a commercial wine industry.

{ 2 }

The Rise of the First
American Wine Culture

GEOGRAPHY, CLIMATE, AND HISTORIC PRECEDENT
ENABLE A CALIFORNIA VINEYARD GARDEN OF EDEN

Agricultural entrepreneurs viewed California's statehood in 1850 as an opportunity to expand the size of the nation and open trade gateways to the Far East and as a means to industrialize agriculture and feed the expanding American Industrial Revolution. Wine enthusiasts also saw the state's favorable geography, climate, and historic precedents as a perfect opportunity to finally commercialize wine making and create a true American wine culture.

In a short time, California had lived up to the expectations of agricultural and viticultural prognosticators. By the early twentieth century, geographer M. K. Bennett observed that California had become a model for specialized, commercialized, and mechanized agriculture with high yields per acre.[1] Geographer James Parsons expanded on Bennet's statement: "In the popular mind it is somehow different, a state with distinctive qualities both of the physical environment and of the human spirit which give it a personality of its own."[2] In a short time, the California wine industry became tied to visionary dreams of wine in moderation and in many ways eventually became part of the present-day, highly political, and contentious culture wars of blue- versus red-state politics in a modern version of Gemeinschaft-Gesellschaft battles of the past.[3] Rural communities distinguished themselves with their evangelical beliefs, rugged individualism, and political conservatism and engaged in culture wars with the urban industrial regions filled with multinational immigrants bound to foreign ethnic, religious, cultural, and political beliefs.

THE NEED FOR A GEOGRAPHY AND CLIMATE
CAPABLE OF SUPPORTING WINE GRAPES

Like all agricultural stories, one must begin with geography and climate. The geographic history of California is a complex story deeply set in plate-tectonics theory. Writer John McPhee described how California had been "Assembled" by thousands of violent earthquakes that brought landmasses from far parts of the world to fashion the region. Parts of the original North American lithospheric plate acted as the prow of a ship as it floated on hot mantle and slipped into and acquired other masses of land, creating California as we know it today.[4]

In essence, parts of California slid into place and divided the state into distinct regions with varying microclimates. Inland from the Pacific Ocean stretches a forty-mile eastward mass that butts up to Coast Ranges on the west and then drops down to the flat sea-level region called the Central Valley. Farther eastward, the valley thrusts up to the Sierra Nevada. Millions of years of volcanic eruptions, earthquakes, weathering, and erosion left the state with large swaths of flat, rich soil capable of sustaining commercial agriculture in the Central Valley and coastal regions.[5]

Adding to the land's capacity was the state's classical Mediterranean climate that was enhanced with transverse coastal mountains that funneled maritime cool, moist fogs to their western slopes. This geography and climate produced four regions capable of producing cool-climate premium wine grapes (San Francisco Bay Area, Monterey, Santa Barbara, and Temecula) and two hot, dry grape-growing regions for table, raisin, and wine grapes (Central Valley and Southern California). Californians quickly identified with and promoted a sense of identity with their Golden State, nicknamed for the 1849 Gold Rush, ample sunshine, and fields of golden grain. Over the next few decades, wine became a large part of this new sense of regional identity, and California earned this sense of place, or what wine writer Matt Kramer labeled "somewhereness," based upon its geography and Mediterranean climate that allowed it to become one of the top food and wine producers in the world.[6]

SPANISH WINE CULTURE TRADITIONS

With California's statehood in 1850, the United States acquired a terrain with the geography and climate capable of sustaining large commercial grape growing. But this was not the only thing that excited early wine entrepreneurs. Adding to the optimism was the fact that California had proven its wine-grape prowess for almost a century under Spanish and Mexican rule. Eastern and southern wine growers had carefully watched and studied Spanish wine successes in the western territories and now made plans to apply this knowledge and expertise to the former Spanish colonial empire, where commercial wine making had peaked as soldiers, Franciscan priests, and entrepreneurs had planted European rootstock vineyards in Alta and Baja California. The Spanish had sustained their wine culture with these vineyards, as they fermented wine both for the Catholic Mass and for table use and had provided a commercial trade commodity for presidio, pueblo, and mission settlements. In the 1820s, as ownership had transferred to Mexican rule, the commercial trade of wine had continued.[7] The pre-American region had an established wine industry capable of supporting a local wine culture.

After 1833 mission wine production had shifted, as Mexican secularization laws had stripped priests of the use of their vineyards, and the wine energy in what would become California had shifted to privately owned commercial operations. Over the next fifty years, wine production centered on the pueblo of Los Angeles, where Americans like Joseph Chapman, William Chard, and Lemuel Carpenter established a small wine industry that produced around thirty thousand gallons per year. This dramatically increased when Jean-Louis Vignes, a Bordeaux, France, winemaker, planted vineyards and started commercial production. In 1855 Vignes sold his thriving enterprise to his nephews Jean-Louis and Pierre Sainsevain, who quickly expanded the operation by purchasing grapes and increased yearly production to more than 125,000 gallons of brandy. These viticultural successes on the western side of the North American continent provided hope for the failing east of the Rockies American wine industry.

GRAFTING AMERICAN AND SPANISH VITICULTURAL TRADITIONS

California statehood promoted the marriage of the entire American wine-making experience, East and West, and catapulted California wines to domestic and international markets. At first, the fledgling California wine industry faced competition for American wine supremacy with small wine producers in Missouri, Texas, Florida, Alabama,

South Carolina, Virginia, Oregon, Washington, Ohio, and New York. These older established regions initially felt secure, as California wines made minimal inroads into their markets during the Civil War years. A short time later as California postwar wine successes increased, eastern growers and producers began to feel threatened.

In the midst of limited success in wine production in eastern, southern, and middle America, it would be a flourish of new vineyard plantings in California's Los Angeles basin, Central Valley, and Bay Area that would grow to dominate commercial American wine making. In a short time, the Golden State surpassed Ohio and New York, the previous state leaders of wine production.[8] By the end of the nineteenth century, California led America in wine production by wedding eastern and midwestern wine-growing knowledge with western climate and Spanish traditions.

CALIFORNIA'S FIRST COMMERCIAL WINE SUCCESSES

After achieving statehood California exploded onto the national wine scene as Southern California, specifically the Los Angeles region, led the nation in wine making from the Mission grapes left behind by Spanish colonials.[9] In 1841 Tennessee native Benjamin Wilson established his San Gabriel Valley Lake Vineyard, and by the 1850s he produced fifty thousand gallons of wine from his 160 acres of vineyard. Matthew Keller, Irish immigrant, came to the Los Angeles region in 1851 and planted vineyards on his 13,000-acre Rancho Malibu and produced more than one hundred thousand gallons per year. In a short time, yearly production from the region topped five hundred thousand gallons per year and bested Guernsey County, Ohio, the former US leader.[10]

Wine-grape growing expanded statewide and soon included counties across the state like Los Angeles, Riverside, San Diego, Santa Barbara, Monterey, Napa, Contra Costa, Sacramento, Santa Clara, and Sonoma. Important for the fledgling industry were a series of favorable governmental policies that attracted investors. Wine entrepreneurs and enthusiasts, eager to promote this new industry, convinced the 1859 California Legislature to exclude vineyards from taxation and started a long tradition of assistance to the state's wine and grape-growing industry. Further government support came in 1861 when the state legislature, in conjunction with the State Agricultural Society, pledged to find the "ways and means best adopted to promote the improvement and growth of the grape vine in California."[11]

Agoston Haraszthy quickly moved to take advantage of the state's supportive mood and traveled throughout Europe collecting more than one hundred thousand cuttings for his experimental rootstock nursery in Sonoma that over time supplied California growers with a variety of grapevines. Haraszthy dedicated himself to improving American wine through a process of replanting with better grapes and use of the European model of premium-wine production. California agricultural reports between 1856 and 1862 boasted that the number of grapevines statewide had increased eightfold. Amazingly, this new growth failed to keep up with consumer demands, and between 1858 and 1861 US wine marketers had to import an average of more than five hundred thousand gallons a year to supply the new American wine culture.[12] Those in the industry saw plenty of room for expansion.

The fledgling American wine industry experienced a vicious wine competition among regional industries in Missouri, Ohio, New York, and California. Eastern vintners accused California wineries of selling their wines under counterfeit French and German labels, which was often true.[13] In a counterattack, eastern vintners placed California labels on inferior wines. The competition intensified as low-priced European wines poured into the American market. In an effort to undercut these lower prices, greedy California growers and winemakers picked grapes before they were ripe, ignored sanitary standards, used barrels previously used for other purposes, added sugar to sour grapes, used free-run juice for white wine and bitter low-quality second-press juice with skin and seed residue for red wines, and sent wine to market before it was ready.

As an agricultural endeavor, the wine industry followed the Gilded Age farming techniques employed by agricultural and food-processing industries.[14] It was a time when new consumer demands forced wineries to rapidly expand and use new technological advances to drive up their productivity. State and federal governments provided university research and development, and farmers shared their trial-and-error knowledge through conversations, journals and treatises, and the formation of trade associations. Most important, agricultural government planners and legislators produced landmark legislation and policies that supported the nation's Gilded Age movement toward an industrial and agribusiness economy. The Morrill Act, Homestead Act, Hatch Act, and creation of the United States Department of Agriculture made farming a top priority for government officials. Despite the new governmental assistance, many in the California industry worried about how quality and brand issues would hurt their sales. In response, the state legislature passed the 1862 Wine Adulteration Act to enforce truth in labeling. The spirit of agricultural innovation, long characteristic of wine making, was intense in California.

A new generation of reputable California winemakers and industry leaders promoted a series of acts, educational programs, and associations aimed at improving grape and wine quality. Leaders like George Husmann, Missouri viticulturist, nurseryman, writer, and professor of horticulture, gained state prominence and promoted increased home consumption to combat intemperance. Husmann's 1863 book, *An Essay on the Culture of the Grape in the Great West,* and his 1866 publication, *The Cultivation of the Native Grape, and Manufacture of American Wines,* addressed the problems of grape growing in the new western regions. Husmann so believed in the possibilities of California that in 1883 he planted a vineyard in the Napa Valley.

Joining the new California industry, Theodore Hilgard, an Illinois lawyer and judge, came to California with his son Eugene and became the champion of increasing quality through science, or as he believed, through "rational winery practice."[15] The junior Hilgard later became a professor of agriculture and viticulture at the University of California, Berkeley, and eventually became the director of the College of Agriculture Experiment Stations. These men and others recognized the advantages of the region with its moderate climate, close proximity to urban populations, and trade opportunities through the San Francisco Bay.[16]

Further commitment to improve the industry with science and technology continued throughout the 1880s and led to what wine historian Paul Lukacs refers to as the "machines in the garden."[17] Grape growers, with the help of the University of California staff, authorized a Board of State Viticultural Commissions in 1880 and moved the industry to depend heavily upon scientific grape growing researched by the University of California. The university also promoted wine-production practices that produced wines that consumers would embrace. Further enhancing the new industry was the commission's promotion of marketing California wine at state, US, and world expositions and fairs.

Yet in a strange twist of events, it would be plant pests and disease that helped rectify much of the state's quality issue. The root louse *Phylloxera,* which favored California winemakers briefly in the 1870s by destroying virtually all French vines, destroyed California's vineyards in the 1880s, and a short time later Anaheim disease, a virus spread by leaf-hopper insects, devastated the Los Angeles industry. In the long run, it would be pests and disease that purged the state of its inferior Mission grape and allowed forward-thinking wine entrepreneurs to replant with more favorable European wine-grape varietals. The pestilence also acted as a mechanism to rid the state of many marginal and disreputable growers and producers. As an end result, Los Angeles's industry shifted its entrepreneurial energy to the northern part of the state, where cool-climate premium wine grapes could flourish. The few remaining vineyards in the state's southern region shifted to table grapes, raisins, and grapes for sweet after-dinner wines.[18] During this period, the state's winegrowers, like nineteenth-century American farmers in general, also experienced economic spikes and peaks brought on by three depressions that flattened the national economy near the middle of each of the post–Civil War decades.

NORTHERN ITALIANS HELP BUILD
THE CALIFORNIA WINE INDUSTRY

Shortly after approving the United States Constitution, the first Congress passed the Naturalization Act of 1790. This first immigration and citizenship law established the policy of welcoming any free white person, of good character, and offered citizenship to those who had lived in the United States for more than two years. For the period between the War of 1812 and the Civil War, the system provided opportunities for waves of immigrants from Western Europe. Irish citizens escaping abject poverty accounted for an estimated one-third of all immigrants and were joined by almost eight million Germans escaping civil unrest and unemployment. A smaller group of about twenty-five thousand northern Italians sought refuge in America, as they fled the political unrest of the 1860s Risorgimento (Italian unification). By the mid-1870s, more than fifty thousand northern Italians called America home, and many brought their wine culture to the new state of California.

Italians played a key role in making California a successful wine region, and by the latter half of the nineteenth century the first wave of northern Italian immigrants had found a means to continue their food and wine ways in their new homeland.[19] A century later, John A. De Luca, faced with neoprohibitionist threats, capitalized on this tradition by highlighting the need for acceptance of the ethnic tradition of wine with meals as a moderating force for a civil society.

In 1881 a group of wealthy San Francisco Italians formed a corporation and offered shares in the Sonoma County Italian Swiss Agricultural Colony (ISC). Initial phases of the business plan included buying land, planting vineyards, and selling grapes to established wineries. The ranks of the wealthy northern Italian investors included Pietro Rossi (Italian agricultural scientist and largest stockholder), Andrea Sbarboro (grocer and founder of an Italian savings and loan), Dr. Paolo de Vecchi (San Francisco surgeon), Mark J. Fontana (California Fruit Canners' Association), Baptista Frapolli (wine merchant), and Dr. Giuseppe Ollino (medical doctor). Sbarboro led the operation and sold three hundred thousand shares of common stock to Italian and Swiss employees, who financed their shares through voluntary installment payments from their payroll earnings.

Problems occurred when skeptical immigrant workers failed to participate in the financing scheme, thus forcing the organizers to enter the wine-making business earlier than expected. Rossi's twin sons, Edmund and Robert, then traveled throughout Europe gathering grape-growing and wine-making techniques and, upon their return, studied enology at the University of California, Berkeley. The colony undertaking faltered in 1885 when the Foran Act made the padrone system illegal and stopped the importation of contracted Italian workers, who exchanged a portion of their wages for passage to America. Despite the loss of immigrant shareholders, the ISC became California's largest wine producer by the start of the twentieth century.[20]

Over the decade of the 1890s, additional Italian immigrants prospered in the new industry. Names like Victor Sioli, Giuseppe Franzia, Edoardo Seghesio, Vincent Picchetti, Felix Salmina, John Battista Cella, G. B. Vincini, Andrew Mattei, Antonio Forni, and Giovanni Foppiano opened successful wineries. More important, the destruction of the Los Angeles wine industry by Pierce's disease drove table-wine production northward to the Bay Area, where Italian-owned wineries flourished. This ethnic dominance of the industry encouraged another wave of Italian wine men that included Samuele Sebastiani, Louis Pagani, Beniamino Cribari, Agostino Martini, Julius Nervo, Lorenzo and Rena Nerelli, Secundo Guasti, Enrico D'Agostini, Santo Cambianica, Demetrio Papagni, Raffaello Petri, Joseph Digardi, and Adolph Parducci.

By the end of the nineteenth century, California produced fifty million gallons of wine yearly, about 88 percent of the nation's total wine production, and the Italian presence in the industry was undeniable.[21] Italian families controlled the lion's share of vineyard land in Sonoma, Napa, Contra Costa, and Santa Clara Counties.[22]

GILDED AGE WINE

By the late 1880s, the successful marriage of America's East and West wine-making knowledge, coupled with California's perfect grape-growing geography and climatic conditions, resulted in the establishment of a sustainable American commercial wine industry. This success prompted California wine entrepreneurs to emulate Gilded Age agriculture and embark on a series of mergers, consolidations, and vertical integration schemes to expand the industry.[23] In a short time, a few businessmen controlled the wine market and produced more than 80 percent of all American wine. They increased production through a wine-by-design technique that included mechanization, as well as use of new scientific knowledge and technology, and they adopted new industrial practices for processing, distribution, and marketing.[24] As a result, wine became an industrial product and followed the same restrictions and processes as most industrial products of the time. As these wine businessmen grew in wealth and power, they nudged both the University of California and the California state legislature to assist them in fighting pests and diseases and lobbied for government-friendly policies that would increase their marketing capacity.

The era of corporate wine production and distribution began in earnest in 1894 when seven well-financed San Francisco wine merchants founded the California Wine Association (CWA). Their goal was to vertically integrate the entire wine-supply chain from vine to store and produce bulk consistent-quality wines in enough quantity to supply national and international markets. The impetus for the organization came from Percy Morgan, an English accountant and financier, who created the monopoly to control supplies and stabilize fluctuating prices. As the director of the CWA, Morgan cared little about who drank, or about why they drank wine, as long as members profited. Under his leadership, the California industry quickly moved from local to national and international markets, and as a result many in the Gilded Age elevated Morgan to the status of "Captain of the Wine Industry."[25] Over the next few years the organization grew to include more than fifty wineries that in 1895 produced eighteen million gallons which grew to twenty-three million in 1900, then to thirty-one million in 1905, and topped out at forty-five million gallons in 1910. The CWA grew to control two-thirds of the state's total wine production, and wine historian Paul Lukacs credits Morgan with "the introduction of wine as a manufactured commercial product, one with a consistent character and brand identity in the marketplace."[26] Wine consumption in moderation for all Americans was finally a possibility.

Most California winemakers started the twentieth century with high hopes, as national wine consumption grew to one-third gallon per capita per annum, and California's yearly production of fifty million gallons supplied 88 percent of the nation's total production.[27] But the trend toward bigger-is-better and monopolized markets faced a new problem, as anti-alcohol forces blamed corporate alcohol producers for most of the ills in American society. As the industry became corporate, wine's image as a beverage of moderation came under attack.

By the early twentieth century, the California wine industry seemed positioned to become a major player in the domestic and global wine market, and entrepreneurs and large industrial corporations moved to provide wine for their seemingly ever-expanding market. But over the next three decades, vintners saw their product become illegal and then become legal again with limitations, a depression, and the conversion of the American economy from peace to wartime status. Although the industry initially suffered, it slowly adapted and rebuilt to a point of readiness for the Cold War consumer era.

After more than two hundred years of failures and mixed results, the American commercial wine industry had finally found a perfect home, and the nation had its first American wine culture. Government leaders who believed in the power of wine to moderate the nation's drinking openly supported the new wine industry as a means to build an American wine culture. Yet historian Erica Hannickel warned of a dark side to this new wine commercialism. She described the California viticultural industry as having a boosterish narrative guided by pastoral Republicanism and Manifest Destiny that left most people with a flawed version of American industrial agriculture, a version without the warts of environmental degradation, racism, capitalist greed, class hierarchy, and anti-alcohol believers. She argued that California's capitalist wine culture had "acted as both a civilizing practice and one that strengthened Americas' grip on the continent" and that "viticulture was thus part of the elaborate, ever-evolving ideology that sanctioned imperial growth in the nineteenth- and twentieth-centuries."[28] Through it all, California agriculture had become a capitalist story with deep ties to consumerism, industrial food, California cuisine, American wine culture, and contentious state and national alcohol policies.

{3}

Loss and Rebirth of an
American Wine Culture

PROHIBITION

At the beginning of the twentieth century, many American citizens bemoaned what they felt was the societal degradation caused by alcohol. To conquer this new national enemy, they set out to wage war on large alcohol-related corporations and usher in the great social experiment of Prohibition. The resolute anti-alcohol forces abandoned the idea that wine could be used as a drink of moderation and lumped all fermented spirits with distilled alcohol. In 1913, after decades of protesting, the temperance movement, in cooperation with women's groups, religious leaders, and anti-immigration groups, pressured states to design and enforce their own alcohol commerce laws. By 1919 thirty-three states prohibited or restricted alcohol. Throughout California, wineries and grape growers faced this barrier by seeking new innovative approaches to legally continue to grow wine grapes, make wine, and profitably market their product.

With the passage of the War Prohibition Act of 1918, designed by agricultural officials to save foodstuffs for World War I, many feared the possibility of draconian restrictions or prohibitions on all alcohol. Their worst fears materialized in 1919, as Congress overrode the veto of President Woodrow Wilson and passed the Eighteenth Amendment that prohibited the manufacture, sale, or transportation of intoxicating liquors within the United States. Assisted by the Anti-Saloon League, now known as the American Council on Addiction and Alcohol Problems, Minnesota congressman Andrew J. Volstead quickly followed up with the Volstead Act that provided enforcement provisions for the new amendment. The sudden change in the nation's alcohol policy stunned American wine drinkers and producers. Over the long haul, for better or worse, Prohibition moderated American alcohol consumption, ended an era of working-class saloons, served as one of the more successful

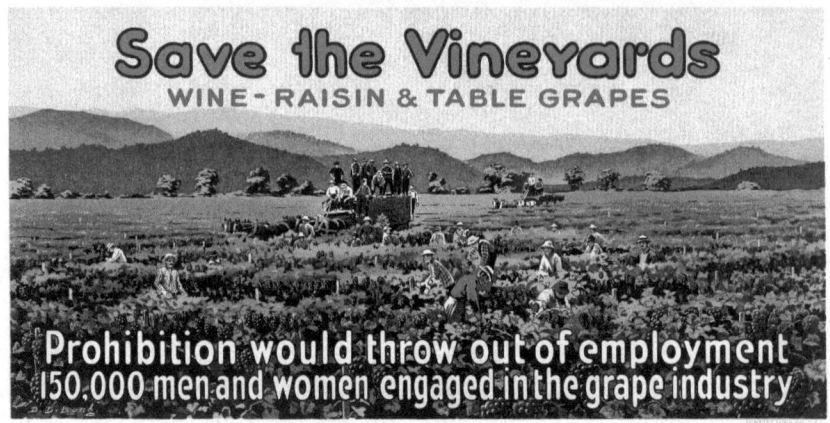

1919 Save the Vineyards poster, part of the culture war to save wine as a drink of moderation and part of a healthy diet. Courtesy of the University of California, Berkeley, Bancroft Library.

alliances of upper and middle classes to legislate morals and habits, and in a strange twist increased wine consumption.[1]

As the nation officially turned to prohibiting alcohol, America's wine culture faced a new problem. Throughout most of America's history, citizens and politicians alike praised the idea that wine was a moderating drink. With Prohibition's anti-alcohol national policy, they had to scramble to help modify laws so as to allow for exceptions for the production, consumption, and distribution of wine. This proved to be the saving grace for Italian winemakers, as the exception provision allowed families to make up to two hundred gallons of wine per year for personal use. In order to meet the grape needs of home winemakers, many Italian American grape growers and vintners assumed a leadership role in keeping the industry alive by increasing their vineyard holdings or grafting existing vineyards to varietals easily transported to families in eastern markets. By 1923 entrepreneurial vintners like Edmund and Robert Rossi shipped grapes and grape juices eastward for home winemakers, who each year crushed more than fifty-five thousand railroad cars of California grapes.[2] Confident that Prohibition would eventually end, the Rossi brothers stored two hundred thousand gallons of dry red table wine so as to be ready for repeal. They also purchased the Asti vineyards of the Italian Swiss Colony in what Edmund Rossi later referred to as "banking on repeal."[3] With commercial wine making illegal, many grape growers like the Petri family and the DiGiorgio enterprise profited by selling wine grapes.[4]

Besides people using their grapes to make wine at home legally, creative wine persons found many ways to keep their vineyards profitable. Limited wine production for vinegar, sacramental wine, medicinal wines, industrial alcohol, cream of tartar, and flavorings could be secured with permits issued by the federal government. As a result, wine production jumped from an estimated 50 million gallons per year before Prohibition to 76.5 million gallons per year during Prohibition. Herbert Hoover's Wickersham Commission counted more than forty-five thousand legal permits in California alone and in 1931 concluded that "it appears to be the policy of the government not to interfere with it."[5] Per capita national wine consumption climbed from 0.47 gallons per year in the years before Prohibition to 0.64 gallons during Prohibition and showed some persistence in that increase at 0.53 gallons per year right after repeal.

While wine consumption increased, the commercial wine industry collapsed, as bonded American wineries fell from 700 in 1919 to fewer than 140 in 1932. Realizing that Prohibition had killed the formal wine industry, vineyardists sought stability for their remaining grape market, and in 1926 anxious California grape growers formed the California Vineyardists Association to stabilize grape prices, secure markets, and enhance distribution systems. Within a short time, they had signed up more than 750 of the state's grape growers. The association proved to be very successful and resulted in red-grape prices jumping from twenty-five dollars to more than eighty-two dollars per ton. As part of their adaptability, many growers grafted their grapevines over to thick-skinned varieties that could be easily transported to eastern home winemakers. Over time this proved to be detrimental to the industry, as it left much of the vineyard land planted with inferior varieties that produced poor-quality sweet and dry table wines that drinkers learned to tolerate. As a result, many premium vineyards and wineries fell into disuse, and for many Americans wine became just another form of alcoholic beverage.

REPEAL

Good news came as the 1932 presidential election left no doubt that most of the nation favored repeal of the Prohibition Amendment. In order to get support for the Repeal Amendment, President Franklin D. Roosevelt promised state legislators and governors that they would receive alcohol excise taxes to help bolster state budgets emptied by the Great Depression. After the election, a wet California Legislature, a wet US Congress, and a newly elected wet president argued for easing up on Prohibition enforcement until the amendment was ratified. Through an executive order, Roosevelt approved both 3.2 percent beer and a 12 to 14 percent watered-down, sweet, carbonated sacramental wine and gave doctors the right to prescribe wine. As a result, wine production tripled between 1932 and 1933.[6] The best Christmas present for the wine industry came on December 5, 1933, as the nation officially repealed Prohibition.[7]

Prohibition had negatively affected wine drinking in America. In 1900 the nation's wine consumers had preferred dry table wines by a two-to-one ratio, and as a result of Prohibition consumers now drank sweet high-alcohol wines by a four-to-one margin. By 1933 the American demand for high-alcohol wine resulted in a doubling of sweet-wine producers in the hot Central Valley of California. This preference for sweet wines continued throughout World War II and peaked at a high of three out of every four bottles of wine produced.[8]

This new trend worried anti-alcohol forces. After losing the Prohibition Amendment, they now doubled down on their quest to stop what they saw as the ravages of alcohol on the social fabric of the nation. The new statistics also worried producers from the international wine community who sought to reestablish their exports to America that had been devastated by World War I, so much so that even Benito Mussolini, infamous Italian premier and dictator, warned winemakers in 1932 that after Prohibition, Americans would have to reeducate themselves so as to be able to "enjoy the noble and delicate pleasure afforded by light, tasty, and refined wines."[9]

But the real blow to the wine industry was that the Repeal Amendment permitted each state to establish its own liquor laws. This legal precedent would forge all future growth for the industry. Thus, the wine industry faced forty-eight (later fifty) different sets of regulations for the transport, sale, taxation, license fees, and distribution of wine. Making matters worse was the fact that just after repeal, nineteen states remained dry. In 1966 Mississippi became the last state to repeal the Prohibition Amendment. Also of concern to the industry was the resurgence of anti-alcohol supporters, later labeled neoprohibitionists. This new anti-alcohol movement grew and quickly continued to wage war against alcohol and its marketing and distribution. For the wine industry to again gain global and national prominence, many local, state, and federal neoprohibitionist roadblocks would have to be overcome. One bright hope for the future remained as small, niche, premium wineries not only survived but began to prosper.[10]

The problem became more complicated, as some states allowed the sale of wine in private grocery and liquor stores, while seventeen others created state or municipal monopolies for the sale of alcoholic beverages. To appease pro- and anti-alcohol forces, Rexford Tugwell, the assistant secretary of agriculture, adapted the age-old position that wine could promote alcohol moderation and proposed that wine be exempted from federal taxation. This moderation proposal met the wrath of Missouri congressman Clarence Cannon, member of the House Appropriations Committee and lifelong prohibitionist. For more than thirty years, until his death in 1964, Cannon used his congressional power to block any help for the wine industry. His political clout was so strong that he forced the Department of Agriculture to strike the word *wine* from all of their publications and in essence created a federal policy whereby wine was no longer considered to be an agricultural product.[11]

Things did not look good for the industry, and Leon Adams, a wine historian and writer, believed that by the 1930s, most Americans thought of wine "as a skid-row beverage" and that the industry was interested only in profit margins.[12] The situation worsened as members of the industry became preoccupied with issues of bulk wine versus premium wines, sweet versus dry wines, high alcohol versus low alcohol, and, most important, the question of whether wine is a beverage of moderation. Just as the wine industry moved to rebuild, anti-alcohol forces vowed to continue their quest to restrict alcohol and save Americans from moral deprivation. In a short time, they convinced state and federal politicians to place wine under the jurisdiction of the Bureau of Alcohol, Tobacco, and Firearms (ATF). This political move branded wine as a dangerous substance that needed to be regulated.

In less than one century, America and California had created, lost, and won back a wine industry. The rest of the twentieth century would be spent rebuilding the industry and re-creating an American wine culture. But bad news lay on the horizon as the nation faced the Great Depression and World War II, which hampered the growth of the industry. The real travesty was the fact that the wine industry had fallen under the control of investors and financiers more interested in profits than in maintaining an American wine culture. This attitude directly fed into the arguments proposed by anti-alcohol supporters. Lukacs believed that "large scale commercial wine making obscured wine's essential identity, making it appear to many Americans to be but another form of alcohol."[13] Dreams of wine as a drink of moderation faltered as anti-alcohol forces seized the moment and lumped wine with all alcoholic beverages.

{4}

Rebuilding the Wine Industry

ITALIANS TAKE A POST-REPEAL LEADERSHIP ROLE

As Prohibition ended in 1932, Italian American winemakers quickly proceeded to help rebuild the American wine industry and reestablish an American wine culture. Decades later, John De Luca appropriately labeled this pioneer group the "Phoenix Generation."[1] These Italian wine pioneers of the 1930s included Ernest and Julio Gallo, Adolph Parducci, Rachele Passalacqua, Louis Martini, John Garetto, Domenico Galleano, Sebastiano Luppino, John Gemello, Emilio Bandiera, Anthony Cappello, Giovanni Pedroncelli, Joseph Filippi, Antonio Perelli-Minetti, Emilio Guglielmo, and Cesare Mondavi. The Petri family utilized their Prohibition profits from their Petri Cigar Company to purchase St. Helena and Calistoga cooperatives and contracted their bulk wines to Gallo.[2] In another scenario, Norbert and Edmund Mirassou, who had refused to shift to easily transportable grapes for home wine making, had the enviable position of being able to immediately supply quality wine grapes to the Cribari and Bisceglia wineries.[3] As wine again became a legal commodity, Italians owned 75 percent of Santa Clara wineries, 72 percent of Sonoma wineries, 59 percent of Napa wineries, and 51 percent of the remaining Southern California wineries.[4]

After the repeal of Prohibition, many struggling wine-grape growers concentrated their energy and grape profits into the wine-making side of the industry. To do so, they engaged in business relationships and cooperatives with their fellow Italian American winemakers. As the revitalized Italian Swiss Colony confronted deteriorating facilities, they also faced shortages of cash, grapes, and bulk juice. In order to compensate for these shortcomings, the ISC first partnered with DiGiorgio and his Fruit Industries Incorporated and later with the Seghesio and Sbarboro families. Happy with their new, vertically integrated endeavor, the DiGiorgios purchased the Trocha Winery in 1932 and received wine-making guidance from Horace O. Lanza and Antonio

Wine Institute annual meeting at the Hotel Del Monte, Monterey, California, ca. 1930s. Courtesy of the Wine Institute.

Perelli-Minetti.[5] By the 1950s two of the five largest wineries were co-ops. But in the words of Ernest Gallo, "By its nature, a co-op was primarily a producer, not a marketer."[6]

ORGANIZING TO REBUILD—THE WINE INSTITUTE

After repeal and during the Great Depression, most wineries and grape growers agreed that they needed to form a cooperative organization capable of rebuilding California's wine industry by reeducating consumers and modernizing their vineyards and production facilities. With this goal in mind, industry representatives met on October 19, 1934, in San Francisco to create the California Wine Institute as a promotional organization for both sweet and dry table wines. Initially, thirty-two wineries signed up, and on November 2 they incorporated, elected A. R. Morrow as honorary president, Harry Caddow as secretary, and Jefferson Peyser as the legal officer. A short time later, they established an office at 85 Second Street in San Francisco. The first members represented about half of the state's total production but only a tiny fraction of the state's then 654 bonded wineries. This changed during the

first year, as membership grew to 188 wineries that produced 80 percent of the state's wine. Initial plans included advertising campaigns, educational programs, and lobbying the state and federal government for favorable tax and distribution policies.

Many forward-thinking organizers realized that repeal had not eliminated the fervor of anti-alcohol advocates, who wished to saddle the industry with high taxes and restricted distribution. Their concerns proved correct, and over the next few years the institute battled trade restrictions imposed by numerous dry states.[7] In an effort to expand their markets, the institute's leadership recognized that they needed to win over congressional and state legislators, and to achieve this they planned to first reeducate Americans, who would then in turn pressure legislators for policies and laws that would return wine to its historic image as a food, medicine, and drink of moderation.[8] Their initial attempts to acquire government assistance failed, as conservatives in Congress and the California Legislature opposed the use of any government funds to support an alcohol industry. To help find a way around the problem, the California Chamber of Commerce brought together winemakers, grape growers, glassmakers, barrel makers, machinery manufacturers, and service industries to strategize a plausible recovery plan.

Overall, most wine businesspeople agreed that the solution involved some means of having the industry support its own costs through assessments from all wineries. After seeking legal and political advice, they investigated bringing the wine industry into the recently passed California Marketing Act. The 1937 act required 65 percent of the growers of a specific crop to ask for a "marketing order" that required all California growers of that crop to pay dues according to their production. This seemed doable to the Wine Institute leadership, and in 1937 they fanned out across the state to sign up supporters. By 1938 almost 90 percent of California's wine producers had joined the organization, and in a short time they initiated the Wine Marketing Act. To oversee the program, the California Department of Agriculture selected a Wine Advisory Board that in turn hired the Wine Institute to implement publicity and educational programs. Under the program, the Wine Institute hired the Walter Thompson Agency to develop a national campaign that included advertising in publications, billboards, car cards, shelf cards, and a series of national wine weeks.

Part of the new program included measures to assure consumers of the quality of California wines. Ninety members of the Wine Institute met at the University of California, Berkeley, College of Agriculture and worked with university staff to promote federal legislation to address the problems of adulterated, misbranded, and low-quality wines. Members also agreed that the institute would start an educational campaign to revitalize the image of wine as a food and temperance beverage.[9] Further discussions continued at the San Jose, California, Sainte Claire Hotel, where members strategized possibilities for expansion into eastern markets, wine-quality standards, wine taxes, and trade practices.[10]

The campaign to increase quality standards continued, and in 1938 the institute petitioned the Federal Alcohol Administration to help California wineries set quality standards. The agency agreed, and for the first time in history the federal government required that all California wines had to be 100 percent made in California and conform to the strict standards of the California Department of Public Health.[11]

With a wine-quality standard in place, the Wine Institute then turned its energies toward its 1939 $2 million national advertising campaign. Their goal was to advertise wine as "another food in the home, stimulating good fellowship and good digestion, and adding to the culture and neighborliness of the American people." At the institute's sixth annual wine-industry conference in San Francisco, guest speaker John Boettiger, *Seattle Post Intelligencer* editor and son-in-law of President Franklin Roosevelt, lifted attendees' spirits with his prediction that "incredible fortune" awaited them. But he then added the caveat that they would have to do more to "standardize their products and protect the good name of California wine."[12] The new marketing programs proved successful, and in 1939 wine sales jumped to 64,560,000 gallons, surpassing the 1938 high of 55,000,000 gallons.[13] Over the next few years, marketing and educational campaigns continued to be successful, and in 1941 the California industry sold 89,237,000 gallons of wine.

While successful as a program, the advertising campaign had the by-product of creating a fissure in membership expectations. Fortified sweet-wine producers feared that the campaign's moderation message favored the table-wine members, and they were right. Just after Prohibition, consumers preferred fortified sweet wine over dry table wine by a four-to-one margin. The success of the dry table-wine marketing campaigns became apparent in 1937 when sales figures showed a 5-million-gallon yearly increase in sales of table wine. This news prompted sweet-wine producers to challenge the institute's promotional programs.[14] Trepidations about a continued plummet led sweet-wine producers to protest a University of California enology program designed to increase research and technical assistance for table-wine producers.[15] Over the next two decades, the battle between dry- and sweet-wine advocates festered.

As a founding member of the Wine Institute, Leon Adams used his skills as a journalist, publicist, and historian to help transition the American palate from sweet to table wines as he proselytized for the merits of wine as an everyday beverage. In an oral history Adams reflected:

> My purpose, and I should really underline this, is to increase the consumption of wine in the American diet. I point out that wherever wine growing has come into existence, the per-capita consumption of table wine promptly increases. We have to civilize drinking

in all of America, not just in California. Americans need only to realize that table wine grows on a grapevine, and that it is a civilizing and civilized beverage. When they recognize the agricultural side of this, when they recognize that wine is food, an agricultural product, they cease to avoid table wine.[16]

Darrell Corti, a Sacramento grocer and gourmet, remembered a conversation with Adams where he proclaimed, "There's got to be two kinds of wine: everyday wine and Sunday wine and wine should cost not more than a gallon of milk."[17]

To accomplish his wine dreams, Adams advocated for small family wineries to produce these dry table wines suitable for meal consumption. To assist in advancing his cause, he enlisted the help of past state legislator and Wine Institute attorney Jefferson Peyser in lobbying for a bill to assist small "family wine-farms." Peyser had little knowledge of the industry and zero experience with wine, but he inherently understood the problems faced by the wine industry. In his words:

> I never knew anything about a grape except what I saw in the grocery store, and we of course never drank wine. That's one of the problems the wine industry had for many years, you see—my generation was the Prohibition era, so all we drank was bootleg whiskey and bathtub gin and things of this nature. That's what took so long to educate our generation to the use of wines, because our generation had completely escaped that. I don't know that it [wine] had any poor image, I don't even know that it had an image.[18]

In 1939 Adams and Maynard Amerine, a University, California, Davis, enology professor, upped their table wine support by forming the Society of Medical Friends of Wine. The organization brought together doctors and dentists who were interested in wine and wanted to support the medical benefits of dry table wine as a moderating food beverage.

WORLD WAR II

Despite these attempts to revive the wine industry, many winemakers still worried about how the nation's more than 380 reopening wineries would overcome cooperage shortages, consumers short on cash, different state wine regulations, outdated equipment, and poor-quality vineyards. Everyone was excited about the new Wine Institute measures that kick-started the rebirth of the wine industry, but in the long run it would be World War II and the subsequent Cold War consumer frenzy that reenergized the industry and boosted the demand for California premium wine.

The post-Depression industry had relied heavily on bulk wine sales that made up 80 percent of all wine produced. This system favored large growers, large wineries, and cooperatives. To survive, most smaller wineries adapted the advice of Dr. Frederic Bioletti, a viticulture professor at the University of California, to establish their market share by emphasizing the sale of smaller quantities of more expensive premium wines. Thus, over the next few decades, innovators like André Tschelistcheff, Carl Bundschu, the Beringer Brothers, Roy Raymond, Louis Martini, and Robert Mondavi established better vineyard practices, planted superior grape varietals, and utilized new technology. Their pioneer efforts set high standards and encouraged others to adopt the new premium-wine scheme.

But larger problems loomed on the horizon, as marketing for small and large wineries faltered under the repeal distribution policies that attacked the concept of "family farm-wineries."[19] These post-Prohibition discussions focused on whether wine was a food or an alcoholic beverage and on whether wines, grown and produced on family wine farms, could be sold at their source like any other agricultural product. To resolve the issue, California state policymakers moved to allow small family-operated wine farms to promote themselves through tasting rooms and direct retail sales. This worked for the California market but set the stage for a future national confrontation over direct sales across state lines, as each state developed its own post-repeal wine-marketing policies and regulations.

World War II had crippled the French, Italian, and Spanish industries and thus blocked most of their international wine trade with the United States. The upside is that this trade disaster provided the opportunity for California wineries to dominate the domestic premium-wine market. But in order to achieve this goal, California wineries had to seriously increase their vineyard acreage and shift to higher-quality premium grapes and wines.

America's international role began as early as 1936, when the

Federal Alcohol Administration Requirement #4 mandated certified labels for varietal bottled wines to make them more competitive with European wines. The mandate also required wines to contain 51 percent of a grape variety in order to carry the name of the grape varietal on the label. This action forced many in the industry to begin switching from bulk blended wines to bottled wines. The adjustment continued in 1943 when the War Production Board converted the last of seven hundred bulk wine-tank railroad cars for use in the war effort, forcing wineries to bottle more wine at the local level. Further support for the premium-wine industry came with the government wartime purchases of raisin grapes, which made up 54 percent of all grapes crushed for wine, in an effort to preserve foodstuffs for the war effort. This new policy also mandated that only premium wine grapes could be used in bottled wines. Together these actions greatly increased wine quality throughout the industry.

Over time World War II served as the vehicle to help the industry develop national brands, increase acreage of premium grapes, and establish the practice of at-winery bottling, and it brought about a massive influx of capital for modernization. The reversal from bulk and sweet wines to bottled dry wine had begun, and in 1944 Ed Rossi, who was president of the Italian Swiss Colony, predicted that sales of dry wine would surpass those of sweet wine.[20] His prediction would become reality over the next two decades.

ITALIANS IN THE POST–WORLD WAR II INDUSTRY

It would be during the post–World War II era of rebuilding and consolidating that Italian American prominence in the wine industry waned, with the exception of Gallo wines, which topped all producers. Second-generation French wine man Philo Biane purchased the Southern California Brookside Winery in the 1960s and marketed and distributed wine through sixteen winery tasting rooms. Brookside used grapes from its five-thousand-acre Italian Vineyard Company to make wine and in the process created the state's largest retail wine business.[21]

As the post–World War II economy created high levels of disposable income, many wealthier drinkers turned to premium wines produced by niche wineries that were owned and operated by a cadre of investors, doctors, dentists, lawyers, and business executives. Despite a new generation of primarily non-Italian wineries, many new American

Italians, like Angelo Papagni, Louis and Andrea Bartolucci, Francis Ford Coppola, Mitch Cosentino, Richard and Paula Crescini, and Don and Rhonda Carano, entered the marketplace. In 1966 Robert Mondavi moved to expand his winery marketing to include mid-level and premium wines, and in a short time his winery became the seventh-largest American winery.

Italian American influence in the wine industry did not end there. In the late nineteenth century, Giuseppe Franzia brought wine to Stockton, California, and in the twentieth century his grandson Fred Franzia revolutionized wine drinking with the introduction of low-priced table wines under the Charles Shaw label (Two-Buck Chuck) for exclusive sale at Trader Joe's stores. Gallo continued its growth, and by the late 1990s the third- and fourth-generation members of the Gallo families controlled one-third of the state's production; became the world's second-largest wine corporation; won the "International Winery of the Year" in 1998, 2001, and 2002; and also won the prestigious Premio Gran Vinitaly.

Italian immigrants had successfully brought their wine culture to the American table. These wine immigrants also acted as business-savvy entrepreneurs and adapted to new grapes, different wines, and new vintibusiness structures in order to build a prominent California wine industry. Yet in another strange turn of the spigot, both Gallo and Mondavi went one step further by exporting and importing American and Italian wines. Like countless Italian immigrants, Italian American wines returned to Italy, changed by their stay in "La Merica."

In the 1980s, Lamberto Frescobaldi, whose wine-making family extended back seven centuries in northern Italy, attended the University of California, Davis's enology and viticulture program and upon his return to Tuscany partnered with the Mondavi family, who had recently purchased the family's Luce della Vite property. *Wine Spectator* magazine touted this as bringing "New World innovation and quality to this family's wine tradition."[22] Luca Currado Vietti, multi-generational Piedmont Italian winemaker, worked in California's Simi Winery, Opus One, and Long Vineyards so he could learn new techniques to reinvigorate his family's winery. At the same time, Mondavi marketed wine in Italy, and in 2003 American tourists at the prestigious Rome Hotel Cavalieri could find seventeen Robert Mondavi Winery Chardonnay, Sauvignon Blanc, Cabernet Sauvignon, Pino

Nero, Opus One, and Merlot wines on the hotel's wine list. Gallo chose a different route and imported the Italian wines Ecco Domani and Bella Sera to America.

POSTWAR REBIRTH RESHAPES THE CALIFORNIA WINE INDUSTRY

Initially, the Wine Institute reorganized the industry and started the process of reestablishing a profitable California wine industry. But it would be the Cold War explosive economic growth that would fully rebuild American wineries and the American wine culture. Hopes for a renewed California premium-wine industry were dependent upon being able to capitalize on the expansion plans. Luckily, corporate America came forward to foot the bill.

Federal regulations to preserve grain products for war needs had forced many corporate distilleries to reduce production, and in an attempt to protect their profit margins they diversified their holdings to include winery ownership.[23] In 1942 National Distillers bought the Italian Swiss Colony for $3.7 million, and in 1945 the Schenley Corporation purchased the DiGiorgio family three-thousand-acre Del Vista Wine Company for more than $10 million. Thus began what observers later called the "Whiskey Invasion." By the end of World War II, industrial food corporations and agribusiness had become deeply entrenched in the business fabric of the California wine industry. As a result, during the 1950s and 1960s, the entire industry moved to a bigger-is-better philosophy, and winery numbers fell from a 1936 high of 1,300 to 271 wineries.[24] Corporate distilled-liquor enterprises funded many of the new acquisitions and mergers and provided an influx of cash needed to rebuild the industry. As expected, this corporatism of wine by liquor companies irritated anti-alcohol groups.

In order to achieve profitability, these new corporate wineries expanded their operations through a consolidation surge that over the next decade reshaped the entire California wine scene.[25] This new era of concentration patterned itself after the general agricultural trend toward larger farms, reduced numbers of farmers, new technology, mechanization, and increased efficiency and production. As the trend accelerated, John H. Davis, director of the Moffett Program in Agriculture and Business at the Harvard Business School, addressed the issue in a 1955 Boston speech. In the talk, titled "Business Responsibility and the Market for Farm Products," Davis coined the term *agribusiness* to describe this centralizing process. He predicted that agribusiness embodied the spirit of a "Second Agricultural Revolution" where vertical integration of production and marketing would run business farms of the future.[26] In California the redesigned wine industry wholeheartedly embraced agribusiness.

In a move that appeared to consolidate the industry overnight, large liquor companies in the eastern United States, such as Seagram, Schenley, and National Distillers, made major investments in California wineries, and in the years after World War II they controlled half of the nation's commercial wine production. This led UC, Davis, viticulture and enology professors Maynard Amerine and Vernon Singleton to make a 1965 prediction that "California winemaking would soon be almost entirely in the hands of a few wineries as their numbers diminished and the survivors grew larger."[27]

The bigger-is-better philosophy also resulted in a doubling of wine-grape acreage from 136,758 acres in 1965 to a 1974 high of 322,044 acres. The increase was led not by regional small pioneers but by the eight largest wine companies, which increased their hold over the market from 42 percent in 1947 to 68 percent in 1972.[28] Fortunately, the liquor companies lacked wine-making knowledge and had little patience to rebuild the industry. Thus, over the next decade, many of the state's wineries gradually reverted back to ownership by resident wine-making entrepreneurs. On the plus side, this merger fever had saved the industry with an influx of much-needed capital for expansion and rebuilding after the ravages of Prohibition, the Great Depression, and war.

CRISIS IN THE WINE INSTITUTE

As wineries fell back into the hands of wine businesses, they also joined the process of reeducating American wine consumers about the benefits of lighter premium wines. For many, this seemed like an uphill battle, as American consumption of flavored sweet wines had increased from 100 million gallons in the mid-1940s to 145 million gallons by the mid-1950s. As a means to promote moderation and shift Americans from sweet wines, the University of California and the Wine Institute openly promoted table wines. This sent ripples through the ranks of sweet-wine producers, and in 1953 the CWA, California's third-largest winery (then owned by DiGiorgio Fruit Corporation), resigned from the institute. DiGiorgio believed that their sweet-wine interests were not properly represented.[29] In many ways they were right. In attempts to build an American wine culture, the university and the institute's educational and advertising campaigns had successfully shifted the majority of consumers to drinking dry table wines.

Switching drinkers from sweet high-alcohol wines got another boost in 1958 when federal alcohol regulations legalized lower-alcohol "pop" wines that had enough carbon dioxide to emit a small pop upon opening the bottle. This led the way to the popular light fruit wines of the late 1960s, and by 1971 overall consumer purchases per $1 million included $346 for table wine, up from $301 in 1969.[30]

All of this growth and consolidation further increased the schism between large and small wineries. After World War II, large wineries pushed to abolish bulk wine sales at the retail level, as they moved to bottling all their wine for the post-repeal three-tier distribution system. Adams felt that this move played into the hands of anti-alcohol forces because consumers would lose the agricultural image of wine as they shifted to buying wine in liquor stores. By doing away with the long-time tradition of the "barrel house" (refilling consumer containers at the vineyard winery), consumers could no longer purchase bulk wine at the winery. Adams worried that this would hurt small family-run wineries that sold directly to consumers. He believed that "wineries out in the country would some day be profitable only if they could sell retail."[31] In the long run, this vision helped establish the legal precedent of the family wine-farm selling retail bottled wine directly to consumers from their estate wineries or tasting rooms. Adams felt it was good for the industry if consumers could go to the vineyard, with all its charm and beauty, to taste and purchase wines. This trend gave birth to the present-day profitable wine-tourism industry built upon the French idea of "wine routes."

The sweet-wine versus table-wine and large versus small rifts in the Wine Institute became so intense by 1954 that Adams resigned from the Wine Institute and blamed large wineries and sweet-wine producers for ruining the industry.[32] Unwilling to give up his cause, Adams continued his wine message and in 1958 published *The Commonsense Book of Wine* that promoted table wines. Wine aficionados praised the book, and after several editions it became a must read for all wine lovers. Adams's legacy included credit for helping to found the Wine

Institute, work for legislation recognizing the idea of a family wine-farm, and a clear message for the need to continue what seemed to be wine's never-ending battle to secure its image as a food and beverage of moderation in a civilized society.

After Adams left the institute, the membership squabbles between large and small producers continued. Many of the large producers disliked the organization's generic sales campaigns and designed their own advertising promotions. Dissatisfied small premium-wine producers argued that the institute should assist them in combating the rigorous and expensive paperwork required by the Federal Alcohol Administration. Adding to the tension was the fact that large wineries gained a competitive edge, as consolidation of wineries resulted in 549 bonded wineries in 1940, down from a 1936 high of 733. The simple fact remained that small wineries could not be competitive with large wineries.[33] Many agreed with this observation, and *Los Angeles Times* wine writer Nathan Chroman, who favored small table-wine producers, warned, "Sour grapes to the well-heeled speculators who have dropped the wine industry like a hot potato and left many to hold the bag. The buying of vineyards and wineries should be for those who wish to make wine, not a fortune."[34] These membership tensions intensified over the next two decades.

The move to large wineries increased throughout the 1950s, 1960s, and 1970s as the industry rebuilt and modernized by shifting to table wines produced by efficient corporate wineries that increasingly relied upon science and technology provided by the University of California. Small wineries fell further behind, and many worried about their survival and pushed for more regulations to help level the playing field. Russel S. Munroe, director of the California Department of Alcoholic Beverage Control, addressed the 1957 Wine Institute annual meeting and forewarned members, "I agree with those in the industry who contend that it is essential to preserve the economic health of the industry. I do not agree with those in the industry who appear to rely upon laws and regulations to eliminate competition and to change the working conditions of economic laws under the guise of preservation of the economic health of the industry." At that same meeting, Don W. McColly, president of the Wine Institute, boasted, "Every aspect of the wine industry from grape harvesting to the bottling of finished wine is being mechanized and automated for greater efficiency and lesser costs."[35] It was painfully obvious that large wineries controlled the narrative.

As always, anti-alcohol proponents, soon to be labeled neoprohibitionists, began lobbying state legislatures and Congress to control alcohol policies and tax issues. These efforts scared most members of the wine industry, who remembered the devastation wrought by Prohibition. At the 1951 Wine Institute annual meeting, President Harry Baccigaluppi, general manager of the Italian Vineyard Company, informed the membership of the Treasury Department's goal to increase grape and wine taxes between 150 to 233 percent. Bottom line, this would have meant an increase of grape prices of up to $120 per ton. He warned that this would "threaten the very existence of the statewide grape industry and is greater than any federal taxes ever proposed for any agricultural food product." Despite threats of tax increases, the industry's 1951 production had topped 124,952,000 gallons, up from 117,700,000 gallons in 1949, and the wine market continued to grow.[36] Additional anti-alcohol scrutiny of the industry resulted in complaints about the institute's use of expensive legislative lobbyists in Sacramento. Of the 450 registered California lobbyists, the Wine Institute's Jefferson Peyser was the fifth-highest-funded lobbyist, outdone only by oil companies, the Public Health League, and Pacific Gas & Electric.[37]

Further stress on the membership came as large wineries, many owned and operated by distilled-alcohol corporations like Heublein and Seagram, threatened to leave the organization because they felt they could go it on their own. By the 1970s the rivalries had become untenable, and one of the victims was the Wine Advisory Board. In 1975, as the board prepared for its customary three-year renewal, Governor Jerry Brown informed members that they must revise their arrangement with the Wine Institute. This came as quite a shock, because for almost four decades the institute had done most of the work for the board. Part of the concern centered upon the Wine Institute's use of money for lobbying and political contributions. The Brown administration felt that Marketing Order dues could not be used to lobby legislators or contribute to their campaigns. The new plan also mandated that the board have a nonindustry consumer representative. Additional pressure by the governor came when state auditor general Harvey Rose investigated the institute's use of $49,995 for political contributions in the 1973–74 campaign cycle. Rose, like the governor, believed that political contributions violated the California Marketing Order.[38] Rather than submit to these conditions, the board voted itself out of existence in June 1975, and as a result the Wine Institute lost its major source of revenue.[39]

The resulting chaos prompted more than half of the members and President Harry Serlis to leave the institute. Those who remained firmly believed that the institute could still serve a purpose, and they aggressively searched for a president who could bring a new leadership approach to the failing organization. Karl Wente took a major role in the search and demanded that candidates understand public policy and political processes and be able to anticipate future needs. The institute needed someone to go beyond the old leadership styles of Adams, Peyser, and Serlis and to move in different directions if they were to survive. Years later, De Luca commented that the "Wine Institute did not know what they wanted—they only knew that they needed a problem solver."[40]

{ 5 }

A Perfect Man for the Job

A Faltering Wine Institute Seeks New Leadership

As the Wine Institute faltered, many within the marketing organization realized that they needed a new direction and a new form of leadership. After many decades of simply marketing and branding wine as a drink of moderation, large fractures had developed within the institute's membership. The membership rupture began in the 1950s and resulted in the resignation of Leon Adams and his message of wine as a moderating force for a civil society. Small versus large wineries, sweet versus dry producers, and those wishing to educate American wine drinkers in the ideals of wine moderation internally battled for control of the institute. The end result was the resignation of top leadership, the exit of more than half of the membership, and an organizational move toward specializing in only marketing concerns.

If the institute were to survive, the remaining members had to seek out and hire a new leader with more than marketing skills. They needed a person capable of bringing warring factions together, someone with state and federal political connections and experience, an individual capable of increasing both domestic and international markets, and someone who firmly believed in the concept that wine was a drink of moderation. Finding a person strong in one or two of these areas would be difficult, but to find an individual who could address all the needs would be a daunting task.

Out of the hiring process, one applicant rose to the top of the list. John A. De Luca became the right man at the right time to redirect the Wine Institute while growing the commercial wine industry and securing the idea of an American wine culture. Born of Sicilian immigrant parents, tempered by growing up on the 1930s–1940s streets of the Italian Lower East Side of New York City, De Luca possessed most of the wanted qualities. He was a believer in the power of multicultural experiences and was devoted to the idea of one-on-one interactions, had training as

John A. De Luca, president of the Wine
Institute from 1976 to 2003. Photo
1964. Courtesy of the De Luca Family.

an academic political scientist and experience in the workings of government agencies, had acquired a long list of political contacts, had
gained experience as a mediator of divergent forces, had proven success in growing business opportunities, and, most important, believed
in the idea of the need for an American wine culture. De Luca embodied the leadership skills needed by the disintegrating Wine Institute.

Over the past century, Italian Americans took a leadership role in
helping their new country establish its first successful commercial wine
industry, and additional Italian American *paesani* then helped rebuild
the industry that had been destroyed by Prohibition and stunted by the
Great Depression and World War II. In the last quarter of the twentieth century, it would be the leadership of American Sicilian John A.
De Luca that would lead them through the struggle to save and rebuild
America's fledgling wine culture.

BUILDING A POLITICAL NETWORK

By the late nineteenth century, labor shortages, driven by the American Industrial Revolution, had changed American cities. To meet the

labor needs of the rapidly expanding nation, Gilded Age captains of industry encouraged southern European workers to immigrate to America. This included millions of Italians and Sicilians who brought their culture, tradition, and religion to their new home. They were drawn to America by promises of economic security and the opportunities provided by the American Dream. At the same time, they were pushed out of Sicily by official Italian government policies designed to make Italy the world's leading exporter of people to the New World.[1] Between 1880 and 1914, more than four million Italians relocated to the United States, with more than 84 percent of them coming from southern Italy and Sicily. Most of the Italians in this wave had agrarian backgrounds and once in their new home formed tight-knit communities based on strong family ties, the Roman Catholic Church, fraternal organizations, and cultural pride. Amid this Italian diaspora, the family of John De Luca came to escape poverty and political persecution and sought the American Dream. They embraced their new land and hybridized its culture by incorporating their own traditions. In the words of John De Luca:

> That [immigrant work] ethic was instilled upon me so strongly by my parents, both of whom had great difficulty with the language. My father at times was a writer in Italy. You would not know that strength of his when he spoke in English. And so knowing that, you think in generational terms. The sacrifices that you have are for those who follow, that you're there to prepare for a new life. And even the day-to-day problems all add up to opportunity, and you overcome that for the larger good. It isn't just people coming in and dumping their poor and their huddled masses; they bring extraordinary energy to the society, vitality.[2]

As a young boy in New York's Little Italy, De Luca learned valuable life lessons and mastered the skill of utilizing politics, religion, ethnicity, and the immigrant work ethic to connect with a divergent group of people. He internalized the importance of food and wine to individuals and group identity. His culture and community taught him that the production, distribution, consumption, ritual, and symbolic values of food defined Italian Americans. Food gave Italians the power to explain themselves and financially supported their families and communities in a world filled with cultural and material stress. Historian Simone Cinotto argues that "food emerged as a nation-building symbol among Italian Americans not only because the wide availability of food in America was the prize won by immigrants driven by hunger, but also because food was at the center of consumer choices that articulated the languages of class, race, gender, and generational relations of power."[3] Most important for this story, Italians gave Americans the gift of their cuisine and wine culture.[4]

At Sunday dinners De Luca learned about his heritage and how wine played an important cultural role as a meal beverage. He clearly remembered the first time when the family gave him the mealtime honor of presenting a dinner toast with a glass of wine. This food and wine heritage followed De Luca for life and became the basis for much of the Wine Institute's struggle to present wine as part of the Mediterranean diet and a beverage of moderation. He also remembered the role of wine in the Catholic Mass and how the priests and nuns taught him the value of all people's lives and how to seek common ground with those of differing opinions.

From his parents he learned cultural pride, the need to involve himself in community politics, and the value of an immigrant work ethic. De Luca quickly mastered Sicilian, Italian, and English, and he learned how to communicate across the varied cultures and belief systems in his multicultural neighborhood. Most important, De Luca internalized the idea that both individual and group dreams can exist in a global community if you listen to everyone and keep adjusting and reinforcing your ideas.

De Luca also learned to navigate difficult times with a belief that hard work, perseverance, community service, and dialogue could benefit the entire community. Events from the late 1930s through the first half of the 1940s taught him how to deal with hardship. The Great Depression, his mother's illnesses, his father's internment as an enemy alien, loss of family members killed on both sides of the war, and separation from a brother in Sicily whom he had never met were problems that he faced head-on. For De Luca, the lesson learned from these hard times was that seemingly impossible situations may take time to resolve and that through perseverance, hard work, and determination, positive outcomes can be achieved. To survive in an immigrant multicultural neighborhood, he learned how to blend different belief systems as a way to work for the common good. In his words: "My personal skill of being able to pay attention to others and internalize what they are feeling and more importantly make them feel wanted while always smiling and saying thank you" had made him successful.[5]

De Luca excelled academically and received a tuition waiver from the University of California, Los Angeles. In 1955 he graduated from UCLA with a bachelor's degree in political science and after receiving two fellowships started work on a law degree at Stanford Law School, where he quickly learned the value of creating a professional and social network. At Stanford the law school assigned Warren Minor Christoper of the Los Angeles O'Melveny & Myers law firm, editor of the *Stanford Law Review,* and later secretary of state for President Bill Clinton, as his adviser. When Christopher realized that De Luca could not envision himself as a lawyer, he sponsored him for a Ford Foundation grant to master the Russian language and work on a master's degree in Soviet studies at Harvard University. While at Harvard De Luca attended lectures from Zbigniew Brzezinski, Polish American diplomat and political scientist, who would later serve as a counselor to President Lyndon Johnson and become a national security adviser during the Carter administration. He also met and studied with Dr. Henry Kissinger, who was a faculty member at Harvard's Center for International Affairs. As a proponent of realpolitik, Kissinger played a dominant role in US foreign policy between 1969 and 1977 as the national security adviser and secretary of state under President Richard Nixon. Looking back at this study experience, De Luca recalled that this high-powered opportunity supplied him with "the lessons of mentorship and making connections. [T]he people I get to know today are the people I will do business with in the future."[6]

Upon completion of his master's degree, De Luca returned to Los Angeles to work on a doctorate in international studies at UCLA. With faculty assistance he secured a Fulbright scholarship to study in Rome and research the question of whether Italy could become a communist nation through the ballot box. His graduate experience in Rome led to an opportunity to serve as a guide at the American National Exhibit in Sokolniki Park in Moscow, where he became a firsthand observer and translator for the now famous Kitchen debate between Nixon and Khrushchev.[7] From the experience De Luca recalled learning a valuable lesson on how to react to a situation where an adversary belittles you. The debate taught him that when disparaged, one must dig deep into their psyche and stand up to the bully and never waver from their beliefs and preferred outcome. He learned to meet shows of strength with resolve and determination.

After the exhibition De Luca continued his graduate studies in Rome, and James David Zellerbach, an American businessman and ambassador to Italy (1957–60), enlisted him for advice on the incursion of communism into Italian politics. Zellerbach arranged for De Luca to participate in four debates with Italian communist mayors from northern industrial cities. His stay in Rome as a scholar also provided him the opportunity to attend Pope John XXIII's Vatican Ecumenical Council and interact with members of the church hierarchy as they discussed worldwide reforms. Through all these experiences, De Luca learned how to question, socialize with, and influence political leaders.

Always open to a new adventure, De Luca then accepted an attaché position at the Moscow embassy to be part of the 1961 and 1962 Transport and Plastics traveling exhibit that traveled extensively throughout the USSR.

During his Soviet travels, De Luca learned that the group that controls the political narrative controls the situation. On a visit to the city of Stalingrad he witnessed the de-Stalinization of Russian history as Nikita Khrushchev renamed cities that had been named for or by Joseph Stalin. De Luca quickly understood a lesson about the power of words to drive a cause or revise history to favor a cause. In his words, "It loosens up your whole sense of your history. It's very dramatic the psychology of the community. The way you're defined when you say, 'I live in San Francisco,' or 'I live in New York,' to suddenly have the name taken away."[8] This situation taught him that it is not an easy task to change cultural heritage and that the group that labels a topic holds the power. This lesson served him well decades later as he navigated the political minefield set by anti-alcohol forces as they challenged his historic and cultural traditions.

While in Soviet-controlled Georgia, De Luca made his cultural food value system apparent in several letters that he wrote to his family. In the correspondence he described his beliefs that all people of the world have more in common than what they differ in and that the Russian people loved the idea of American-style democracy and capitalism. He was very impressed by the hospitality, food, the region's wine quality, and how the locals, like Italians, used wine as a moderating meal beverage. John described to his family the high quality of the wines and how he drank his fair "share; but, only at mealtime."[9]

All politicians must learn to face and respond to accusations about their character. For De Luca, this came when Russian ballerinas befriended him, and members of the United States Information Agency alerted John that he was possibly being targeted for secret information in a "honeypot" sting by the Russian women. At a later date, the Georgian foreign minister lodged an official complaint against John to the State Department "for slander against the CCCP." After American counterprotests, the Soviets dropped the charges, and the incident made De Luca a celebrity among the embassy staff.[10] His Russian experiences taught him that one's opponents will throw scandalous personal labels at you in an attempt to diminish your cause or distract listeners from your message. He learned that you have to endure false accusations while standing your ground to defend your key beliefs.

After returning home to UCLA to complete his doctorate, De Luca began lecturing on Soviet life and politics at the University of California campuses at Los Angeles, Riverside, Santa Cruz, and Irvine and accepted a full tenure-track position in international studies at San Francisco State University. Just as the young professor began his career, another set of swift changes moved his life in new directions. He met and proposed to Josephine Titone, whose family also came from Sicily, and accepted a position as a White House Fellow. He decided to move his career to the political realm and set out to make himself known to the power elite. Senator Margaret Chase of Maine, John Gardner (at that time president of the Carnegie Corporation and the Carnegie Foundation for the Advancement of Teaching), First Lady "Lady Bird" Johnson, Supreme Court justice Thurgood Marshall, and Vice President Hubert and Second Lady Muriel Humphrey joined a list of his new friends. Looking back, De Luca marked the White House fellowship as the seminal moment when he realized that he had become an American. In his words, "I knew I was accepted when I was asked to be a White House Fellow at 'la Casa Bianca.'"[11]

As a White House Fellow, De Luca forged close ties with staff and his peers in the first class, and many remained lifelong friends. Initially, because of past experience, training, and security clearances, the program administrators assigned De Luca to the Vietnam Coordinating

Committee under the direction of national security adviser McGeorge Bundy. His assignment was to provide research and white papers for Secretary of State Dean Rusk, Robert Komer of the Central Intelligence Agency, Bundy, and American diplomat Chester Cooper. It was here that De Luca learned how to research and write white papers designed to help decision-makers analyze key facts on a given topic, a skill he would utilize numerous times as leader of the Wine Institute.

As part of his other White House duties, De Luca befriended White House press secretary Bill Moyers and White House special assistants Joseph Califano and Jack Valenti. In later years Valenti and his wife, Mary Margaret, became personal friends of the De Lucas, and John invited him to speak at numerous Wine Institute meetings to address issues like branding, marketing, and political strategies. A decade later, Califano became secretary of health, education, and welfare and worked on initiatives that included tobacco, alcoholism, and the War on Drugs. De Luca remembered conversations with Califano where they discussed wine's role in their heritage and upbringing. De Luca believed that "not just because of me, but because of his background," Califano took a moderate approach to wine issues. He remembered telling him, "You know, you remember the way we drank in the family. You remember we drank in moderation."[12] This relationship proved to be initially critical for De Luca's struggle to rebuild an American wine culture.

All members of this first class of the White House Fellows had the benefit of learning from an amazing group of mentors. For De Luca, people like David Rockefeller, chairman of the board of the White House Fellows, and John Gardner, designer of the program, stood out. Years later when Gardner joined the Stanford Research Institute Board of Directors, De Luca kept in contact with him and his wife, Aida. De Luca saw Gardner as a master political mentor and one of the greatest politicians with whom he had ever worked. For De Luca, he was a role model. "We would have dinner with him just to be exposed to a man like that. He was secretary of health, education, and welfare, and resigned from the Lyndon Johnson administration without a big to-do because of his lack of support for the Vietnam War."[13] For John, one of his greatest honors came in 2012 when he received the John W. Gardner Legacy of Leadership Award. After his one-year fellowship De Luca served as a staff member for Idaho senator Frank

Church. Over time, all these experiences helped him build a network for future professional assistance.

After leaving his one-year fellowship, De Luca enlisted friends Moyers and Valenti to help him find work in the San Francisco Bay Area. The two, with the assistance of Benjamin Swig, a Bay Area real-estate developer and owner of the Fairmont Hotel, found him a staff position on the mayoral election committee of Joseph Alioto, a San Francisco lawyer and business owner. Alioto won the mayoral job with a brilliant campaign and had been impressed with De Luca's work on his behalf. As a reward he offered De Luca the position of chief of staff. This was a great introductory position for someone wanting to leave their mark in local and state politics and rise through the ranks of the Democratic Party. The excitement soured when Richard Nixon's attorney general, John Newton Mitchell, undertook a dirty-tricks campaign aimed at Democratic mayors across the country. In the case of Alioto, the political hit job included disinformation that *Look* magazine used in their September 23, 1969, issue. The magazine claimed that Alioto had business and personal ties to Los Angeles mafioso boss Jimmy "the Weasel" Fratianno. Alioto later sued *Look* for libel and won a $450,000 judgment.[14]

While Alioto defended himself against the powerful accusations against his character, he convinced the Board of Supervisors to create the position of deputy mayor to assist him while he faced time-consuming legal battles. After creating the position, Alioto asked De Luca to step into this role and turned over the day-to-day operation of one of America's largest cities to him.

De Luca quickly relied upon his political science and international studies, academic credentials, political experiences, family, and cultural skills, and he employed past mentors and colleagues to help him run the city of San Francisco. In a meeting with Alioto, the two discussed the Federal Highway 280 project that would destroy natural habitat and watershed lands as it plowed through the center of the Peninsula's Christmas Tree Reservoir area. John suggested that they take the issue directly to President Johnson and arranged for a meeting at the White House. At the Oval Office meeting, President Johnson asked, "What do you want to do, Johnny boy?" and promised to allow San Francisco to select a better route for the interstate that eventually ran 57.5 miles from San Jose to San Francisco. After discussing the freeway

Left to right: San Francisco mayor Joseph Alioto, President Lyndon B. Johnson, and Deputy Mayor John A. De Luca at the White House in 1969. Courtesy of the De Luca Family.

project, Johnson asked his special assistant Joseph Califano "if there were any more apples in the bushel for San Francisco." De Luca suggested that the city needed an arts center and could better utilize some federal lands in and around the city. As a result, San Francisco secured partial funding for the Performing Arts Center, release of Fort Mason lands for an Italian cultural center, and old Fort Baker lands to establish the Golden Gate National Recreation Area. De Luca had mastered the art of the political deal.

As deputy mayor, De Luca's accomplishments grew to include developing the economic future of the Port of San Francisco, helping to oversee the building of the Bay Area Rapid Transit system, and reviving the TransAmerica building project. He also took advantage of his position to begin building President Franklin D. Roosevelt–style political alliances with the African American, Mexican, Chinese, and Italian communities. He then extended his coalition building to include labor unions and the California Teachers Association (CTA). De Luca also used his new political role to appoint Nancy Pelosi to her first political position on the San Francisco Library Commission, and he developed a political alliance with mayoral opponent Dianne Feinstein and her second husband, investment banker Richard C. Blum. This broadened circle of supporters grew to include Hollywood leaders who took notice of the city and featured San Francisco in the 1971 movie *Dirty Harry* with Clint Eastwood and the 1974 film, *The Towering Inferno* with Steve McQueen and Paul Newman, as well as in the television police series *Streets of San Francisco* with Karl Malden and Michael Douglas. As deputy mayor, De Luca oversaw production contracts with the San Francisco Police and Fire Departments, approved shooting locations and logistics, and hosted the stars. As part of the filming of *The Towering Inferno,* De Luca got producer Irwin Allen to purchase a fireboat to use in the film and then to donate it to the city. De Luca had begun building a vast and complex web of political, business, educational, labor, and ethnic coalitions that he would draw upon for future projects.

CATALYST FOR ANOTHER CHANGE

Despite all his accomplishments, De Luca painfully remembered the darker side of his job, especially the 179 days between 1972 and 1973 when seventy-one Bay Area citizens were murdered. De Luca's position included oversight of the Fire and Police Departments, the city's school district, political protest rallies, and strikes. He remembered dealing with the 1969 Zodiac serial killer case, the Zebra Killings, Haight-Asbury hippie protests, the Black Panther Party, Patty Hearst and the Symbionese Liberation Army, anti–Vietnam War demonstrations, the American Indian Movement takeover of Alcatraz Island, and the rise of the gay rights movement. Further job stress came when pay disagreements stalled labor negotiations between the San Francisco Police Department and the City of San Francisco which turned into a violent strike.

As a result of the political turmoil, De Luca began to fear for his and his family's security and lives. He remembered, "One time at the mayor's office Joe was at his desk and I was in a chair next to it, and we were talking. The door swung open, and a man entered with a gun and before anything could happen, Eddie Seraille, one of our three assigned police officers, tackled him and dragged him out and arrested him."[15] At one point, the De Luca family executed evacuation protocols from their home because of personal threats. Death threats and job stress had taken a toll on the De Luca family, and John began to look for work outside the political realm.

NEW BEGINNINGS

As De Luca began the search for a new career, he knew that the University of California was running a search for a new president, and he arranged meetings with eight of the regents to express his interest. He also knew that past colleagues could help him return to the Stanford Research Institute, he knew he qualified for a "last-ditch" position as the city manager of Anaheim, California, a new opening in local government. Many Alioto administration friends also suggested that he take the lifetime position as the San Francisco chief administrative officer.

His wife, Josephine, pined for a more normal life for her family—one without death threats. She had seen an article in the newspaper about Harry Serlis stepping down from the leadership of the Wine Institute. Thinking that the job would be a perfect fit for the family, she forwarded John's résumé, without his knowledge, to David Rockefeller, Bill Moyers, Jack Valenti, and Frank Church and asked each of them to recommend John for the Wine Institute position. Unbeknownst to the De Lucas, the Wine Institute position was a job filled with pitfalls and challenges.

After Josephine's outreach, De Luca got a call from a headhunter asking if he would be interested in an interview for a position at the Wine Institute. He told the recruiter no and explained that the position did not interest him. Two days later, De Luca received a second call asking if he would go to a Napa Valley luncheon hosted by Louis Martini and Joseph Heitz. He again said no and asked the recruiter to stop bothering him. That night at dinner he discussed the offers with Josephine, and she firmly insisted that he call them back and accept the invitation and tell them that he would be bringing his wife. Guests at the luncheon included Louis Martini, Robert Mondavi, Paul Davies, and Joe Heitz. John had no idea that the luncheon had been put together to informally interview him. Little did the interviewers know, but this was the first time that De Luca had ever visited a winery facility or vineyard. De Luca described the luncheon as nice but prolonged as everyone took the time to talk to him. Finally, Josephine looked at him and clued him in that he was being interviewed. A couple of hours later, Heitz asked him if he could recommend him for an interview for the position of president of the Wine Institute.[16]

De Luca agreed to continue talks about what the position entailed. At first he struggled with the idea of working in the wine industry because he could not see any connections between the institute and his training and work experiences. At home Josephine kept reminding him that "it's the wine industry, and it's perfect for your background." Later at a private meeting, Ernest Gallo told De Luca that he did not think he was the best choice because he thought he carried too much negative baggage because of his exposure in the papers dealing with San Francisco's murders, strikes, and community changes. De Luca believed, "They were very skittish about bringing in a person who daily was putting out a dragnet in San Francisco."[17] They were not sure whether they wanted to give De Luca a formal interview.

De Luca decided that he wanted an interview just to prove that he was up to the task. For John, their lack of confidence in his skills "got my juices flowing, I said, 'Okay, you don't want to interview me, I'll make sure that you interview me.' I made sure that people at least had to interview me."[18] With his persistence he got an interview and remembered, "When I went to the interview I thought I was going to meet this group, see them for an hour, and that would be the end of it. But at least I would have had the interview." He believed that it would be the softest interview in his life, because he had absolutely no expectations.

Years later De Luca recalled:

Ed Mirassou told me, he said, "John, let's go through this, but you already have three strikes against you. But let's go through it." And I said, "Perfect for me, Mr. Mirassou. I've got a car waiting, and I've got a police strike that is waiting to be solved." It was a blur. The institute chairman, Bob Ivie, head of Wine Guild, supported me, but there were a number of fine candidates in the mix, and I carried big-city baggage. I never realized until many many years later that one man really stood out, a champion for me, and that was Karl Wente, Phil and Eric's father. And he saw something in me. I'm told he said, "Look, we are facing problems with Rose Bird [then California secretary of agriculture]. We are facing problems in Sacramento. We've got Jerry Brown on our back. We've got the United Farm Workers. This guy is tough; this guy has had to run a city. Look at his background. Maybe instead of a PR guy, he is the kind of guy we should have."[19]

Because of Josephine's constant persistence, John pursued and accepted the job on a conditional six-month contract. He agreed that it would be nice to be in an industry where they could raise their children around what Josephine considered to be "cultured and cosmopolitan people." Looking back, he recounted, "My wife made the best choice for me."[20]

RIGHT MAN AT THE RIGHT TIME

Many in the Wine Institute selected De Luca as their leader because they believed that they needed more than a businessperson to revive the floundering organization. Wine leadership knew they needed someone with political ties and that De Luca as deputy mayor of San Francisco had successfully navigated the city's turbulent times and developed a skill set necessary to resolve the Wine Institute's internal conflicts. They also believed that his experiences in state and national politics, influential contacts, success in building global business deals, background in building coalitions, and upbringing in a wine culture suited their needs. In De Luca's opinion, they were acting out of desperation and figured that they had nothing to lose.

The *Los Angeles Times* declared that De Luca "is taking the helm of the wine industry's trade association at one of its lowest points in its thirty-seven year history."[21] The newspaper article went on to list key items in the turmoil, including the facts that six more wineries had jumped ship, including the second largest, United Vintners; the president had resigned; the state was auditing its books; the Wine Advisory Board had been abolished; and the institute had fired its public relations firm, the Daniel Edelman Agency. The article ended by declaring that De Luca was not an experienced wine man. Then came a devastating discussion between De Luca and Ernest Gallo, who told him, "People do not talk very highly about you coming. I am going to give you six months." John replied, "Give me the ball and let me pitch."[22] Six months later in a conciliatory gesture, Gallo invited Josephine and John to his home for dinner.

De Luca figured that he could do the job. After all, he had a bachelor's, master's, and doctorate degrees from UCLA and Harvard. Add to that the skills learned from years working in embassies, debating communist Italian mayors, experiences with the Soviet people, being a White House Fellow, working on the staff of a US senator, and being deputy mayor of San Francisco. A confident De Luca knew that international responsibilities, religious upbringing, parental modeling, cultural pride, advanced education, political experiences, and hard work had taught him the value of a network and how to persevere and use transparency to accomplish major tasks.

De Luca's personal conclusion was that the Wine Institute situation had become so toxic that they faced imminent failure and that key members made a measured act to reach out to a person with his background. For some, he was a person who could come in without the baggage of past rivalries and bring policy and political experience to the position. For those wishing to expand the industry to global markets, De Luca's background in international affairs made him a top candidate. The Wine Institute decided to move in new directions and take a risk on a nonbusinessman with no wine-industry experience.

{6}

Stabilizing the Wine Institute and Reestablishing a Tarnished American Wine Industry

Essentially, when people ask me, "What is my job description?" I just basically say I'm a teacher. The board of directors meetings, oh, did I drive them crazy because every board of directors meeting I always felt I had a lesson to impart. And I just simply—we had to do "this" today, and we had to do "that" today, and we had to fight this, the commanding heights of strategy. I learned that dealing with Marxism and Leninism. What are the one, two, three most important things that everything else is subordinate to? And so, reading, writing, lecturing, and face-to-face encounters became my method.

—JOHN DE LUCA, "President and CEO of the Wine Institute," oral history

Despite Gallo's six-month warning, De Luca accepted the position of president of the Wine Institute, and in his typical take-charge fashion he hit the ground running. He drew heavily upon his past family, ethnic, religious, academic, and political skills to begin the process of rebuilding the institute. His initial action was to reflect and decide upon a few simple first steps. He believed he got this tactic from his family:

I learned a nonattack skill from my parents. They taught me to show your standards but at the same time seek common ground. They taught me the value of the church. Do not be an empty shell; stand for something, but always look for a way forward and to learn and cooperate. From my father I got the component of strategy . . . where does this fit in in the short term and the long term. I learned by watching my father as president of the Tito Minniti club on how to conduct yourself in hard times like World War II, his internment, the lessons of moving from New York to Los Angeles. He taught me to think things out, to think them through before I act.[1]

De Luca figured that he had little time, and, more important, he believed that he had to be true to himself. In looking back, though, he believed that some of his success was due to his naïveté. John admitted that he did not understand all the nuances between the various factions, and he tried to stay out of their squabbles and navigate a middle road. He reflected, "I came in with a tabula rasa, a clean slate. I wasn't weighed down by the internal battles. I saw my job as one from keeping the organization from collapsing.... When I entered the institute job, I felt that there was no area that was not on the table. I did not endorse one philosophy over another. I saw an opportunity and a challenge at the same time, and I fell back on my instincts and my training."[2]

De Luca also relied upon his religious training and his experiences growing up Catholic on the Lower East Side to help him rebuild the institute. As a PhD student in Rome, John had learned a valuable lesson from Pope John XXIII's Ecumenical Council. He had felt privileged as he met cardinals and archbishops in the church's hierarchy as they discussed worldwide church reforms. As a result of these deliberations, he had internalized the idea that some battles may take a lifetime. More important, De Luca believed that these conversations had taught him "that large institutions can take an introspective look at themselves and revitalize themselves."[3]

The Wine Institute staff had been drastically downsized to include just Harvey Posert doing public relations, Jim Seff as legal counsel, Al Allmendinger covering health and safety, and a few support staff. From the beginning, De Luca wanted to retain the San Francisco office so that he would not have to relocate his family; had close proximity to Stanford, UC, Berkeley, and the University of California, San Francisco; and would be close to industry leaders in Napa and Sonoma Counties. The institute continued to maintain small one-person offices in Fresno and Sacramento, along with a few state contract representatives.

With the support of Posert, he established an initial goal of bringing a collection of different voices to the many problems the institute faced. At his first board of directors meeting at the San Francisco Palace Hotel, John recalled, "The initial meeting had a polite response, but I knew I had a long way to go. The collective attitude seemed to be to wait and see."[4] He then went to the University of California, Davis, and met with Maynard Amerine and other university staff. From there he met with Peyser and Serlis to pick their brains while being sure to not speak negatively about any of the past or present Wine Institute leadership or members. John also began to kindle relationships with the media so as to build an outlet for institute policies and concerns. He credited his White House Fellow mentors Bill Moyers (who was serving as editor and chief correspondent for CBS *Reports*) and Jack Valenti (who held the role of president of the Motion Picture Association of America) for teaching him how to work with the media.

Drawing heavily from his past political experiences, De Luca's first action plan was to travel statewide and visit hundreds of wineries. John maintained:

> Instead of keeping with the old executive leader from above, I went out to the members. I treated it like I was running for office and went out to meet the constituents. There were over one hundred members who had left or were eligible to become members that I visited. . . . I knew that I had to meet all the wine people because that was my training. In Russia and in Rome I went out to meet people. I took the train to small villages and met mayors to debate. It came down to how you conduct yourself not just in business and politics but in life. Do not attack people but try to find areas of mutual understanding.[5]

Meeting the membership became his means of acquainting himself at the ground level with the problems of the industry. During the first few months, De Luca was on the road two or three days a week visiting wineries throughout Fresno, Central Valley, Napa, Sonoma, Mendocino, San Luis Obispo, Monterey, Santa Clara, and Santa Barbara Counties. Besides meeting members individually, he also asked to be invited to their regional meetings. For many disgruntled members, this was the first time that they felt they were recognized and represented. In a short time, his meet-and-greet technique stabilized membership numbers and helped the majority of the membership identify with their new president.

De Luca's past life on the Italian Lower East Side played a large role in his initial acceptance, and his presence excited many in the Italian segment of the California wine industry. One Italian vintner told him, "finalmente uno dei nostri" (finally one of our own).[6] John helped spread a sense of connection through their shared ethnicity by going to their homes, sitting at their tables, walking with them through their vineyards, and making sure to ask them what they personally needed.

In this early, hectic period, Josephine assisted John by reaching out to the women and reinforced the idea of this being a family industry. Her background in her father's wine shop provided a comfort level, as she and John visited many of the vineyards. Through it all, De Luca tried not to be seen as a promoter of any one part of the industry, and in a short time many members grew to trust their new leader and learned to utilize his political connections to communicate with state and federal officials.

In a move to make the organization more transparent and responsive to the large- versus small-winery tensions, De Luca reorganized the Wine Institute Board of Directors. He developed a new, leaner structure that equalized the power of both large and small wineries. By utilizing the idea of the United States Constitution's bicameral legislative body, he persuaded membership to create two types of members on the board of directors. First would be those voting by volume (favors large wineries), and a second voting group would give each member one vote (favors small wineries). He also created a series of specific committees that reported to the board.

In a typical academic approach, De Luca then began to read, research, and analyze the various issues faced by the industry. He had the Wine Institute secretary gather a reading list from as many sources as possible. One of the sources was Leon Adams's *The Wines of America* that reconfirmed to De Luca that the written word carried a lot of weight.[7] Adams had attempted to write about the needs of the industry and believed members needed to discuss issues and develop strategies to educate the American public on the benefits of wine. The book encouraged De Luca to use this approach as a starting point for his leadership:

> I identified with his written approach because as an academic that was the way I was trained. Also trained by my father this way. I believe he believed that I had an honest need to learn from the past to enrich the future. Where we differed is that I had eight years' experience running a large city . . . fire department, police department, Hetch Hetchy, public utilities, airport, port, housing,

homelessness, political unrest in the streets, businesses, and varied political voices. I had experience dealing with the everyday problems of a large organization.[8]

John planned on continuing Adams's struggle to reestablish an American wine culture, but he developed a process that best suited his life experiences and ethnicity. One area he chose for diverging from the Adams approach dealt with the Wine Institute's dealings with corporate distillers like Seagram that had made wineries part of their portfolio. Because of resistance from anti-alcohol forces, Adams felt that any connection with corporate distillers would hurt the chances of building a wine culture based on moderation. De Luca decided to enhance wine's image without degrading distilled spirits and for years refused to attack corporate alcohol makers when he approached health officials, the press, and politicians. He firmly believed that his ethnicity and upbringing would make it easier to argue the historic value of wine as food, medicine, and cultural beverage of moderation without allying with or denigrating distilled spirits.

De Luca's cultural perspectives paralleled the historic reflections of Italian wine pioneer Andrea Sbarboro. In 1908 Sbarboro began to worry about the temperance movement and how it would affect the California wine industry. In response, he put together the short publication *Temperance vs. Prohibition: Important Letters and Data from Our American Consuls, the Clergy, and Other Eminent Men* to spell out how European wine cultures helped moderate drinking habits. In one of the book's typical letters, William Henry Bishop, the American consul, in Palermo, Italy, laid out a case for an American wine culture. He stated, "Here in Sicily, while the people no doubt have their faults, it is a pleasure to witness their general temperate habits. Even the usual place where drinks are publicly sold is most often a *pasticceria,* cake shop, or a pleasant cafe, which the most respectable persons, including ladies and children, may freely enter. There are no screens put up, there is no concealment of the interior or the inmates, for nothing takes place there requiring concealment. The drinking saloon in an offensive sense can hardly be said to exist at all."[9]

After losing the moral high ground of Prohibition, anti-alcohol forces now sought new ways to politically implement and enforce regulatory alcohol taxes, policies, and laws. They may have lost the ability to prohibit booze, but they were determined to continue their battle against what they perceived to be the evils of alcohol. In the early 1980s, Bruce Yandle, dean of the Clemson University College of Business and Behavioral Science, observed this shift in policy direction and labeled the new strategy "Bootleggers and Baptists." In his words: "Politicians need resources in order to get elected. Selected members of the public can gain resources through the political process, and highly organized groups can do that quite handily. The most successful ventures of this sort occur where there is an overarching public concern to be addressed (like the problem of alcohol) whose 'solution' allows resources to be distributed from one group to another."[10]

In this new policy paradigm, economic and moral forces join together in a strange-bedfellows scenario to shape government regulatory policies for both profit and moral persuasion. Yandle's theory on public choice economics drew from the work pioneered by Nobel laureates James M. Buchanan, head of The Center for the Study of Public Choice at George Mason University, and Gordon Tullock, a professor at George Mason University School of Law. The phrase "Bootleggers and Baptists" emanated from the fact that both groups supported the laws requiring the closure of liquor stores on Sundays. Bootleggers increased profits by having their competition closed one day a week, and Baptists felt they were keeping the Sabbath free of demon run.[11]

From his life on New York's Sicilian and Italian Lower East Side, De Luca understood the cultural value of wine as a moderating force. But he also realized that the new post-Prohibition narrative had shifted from morality issues and now emphasized the idea of public health. De Luca understood that he had to retool the institute's arsenal if they were to win the war. He believed that

> [there was an idea] that in new garb, with new techniques and new tactics, [and that] our country was experiencing some of the same fervor, some of the same activism as occurred in the early part of the century. But it was in terms of public health issues rather than, say, the moralism over the fundamentalism of the early part of the century. And, therefore, it put a premium on analysis. It required of us a better perception of what it was that we faced. It made us aware that we should look at the legitimate health issues that were going to be discussed. Therefore, you could not have a blanket

rejection of these subject matters; and we had to be highly skilled in terms of dealing with these subject maters; and we had to be highly skilled in terms of dealing with government and with the media and understand the social issues of our times.[12]

FIRST ACTIONS FOR THE NEW LEADER—LABELS, ADVERTISING, AND PRINCIPLES

As expected, many of the immediate anti-alcohol problems would not wait for De Luca to reunite the membership and revitalize the role of the Wine Institute. Within his first two years, threats from the federal government to enforce new labels and attacks on alcohol advertising threw the industry into another crisis mode. As a response, De Luca successfully rolled out the 1977 Code of Advertising and the 1978 Declaration of Principles.

Labels

Since the 1935 creation of the Federal Alcohol Administration, regulations for labeling and advertising had purposefully been left vague and at the discretion of the administrators, who decided whether manufacturing processes, analysis, guarantees, and scientific or irrelevant matters were likely to mislead consumers. Many complained that over the years, this process had often been arbitrary and restrictive.

Just a few months after De Luca took the helm of the institute, the Bureau of Alcohol, Tobacco, and Firearms (ATF) initiated a proposal to place seals on American wines. The bureau wanted the seals to guarantee consumer information about a wine's vintage and origin. Such labels were common in countries like France, Italy, Germany, and Spain. Yet others warned that these labeling customs had become a "jungle" of restrictive rules. Despite the general fear of the required labeling, many premium California wineries were already using labels with the term *Estate Bottled*.[13]

John and the Wine Institute vehemently opposed the idea on the grounds that it "would lead to the unjustified inference of government approval of a higher quality wine."[14] They also argued that wine would have to be produced under much tighter standards, which would result in higher prices and reduce consumption.[15] Wine Institute counsel Peyser bemoaned the direction of the proposal and argued that "California wine making and labelling rules are currently stricter than the federal code," and he added that California vintners "have a long history of espousing and attaining high standards."[16] In a November 1977 public hearing on the proposed labeling requirement, George Vare Jr., Geyser Peak Winery and chairman of the Wine Institute's committee on laws and regulations, informed officials that the "requirement of the percentage on the label inevitably suggests to the uninformed buyer that higher numbers mean better wines, which in reality may not be the case."[17]

De Luca testified to the hearing board that the California wine industry was in a growth phase and that the majority of its members were small agricultural family operations that grew grapes, made estate wines, and drew more than four million visitors to their facilities yearly. He added that the seal requirements would only hurt their operations and that they favored anti-alcohol supporters wishing to reduce wine sales. As president of the California Agri-Council on International Trade, vice chairman of the Agricultural Action Committee, and a founding member of United California Agriculture, De Luca questioned the overregulation of agriculture, wine in particular, and told the hearing board, "We know it is not ATF's intention to appear to affix a seal of government approval on higher quality wine. But we believe that the marketplace and the consuming public will not let that seal be a neutral symbol."[18]

More than fifty witnesses, including winemakers, educators, entertainers, grape growers, retailers, importers, and consumer advocates, testified before the seven-member panel.[19] After listening to the Wine Institute's testimony, *Los Angeles Times* wine writer Nathan Chroman wrote that he disagreed with the Wine Institute. In his words, "I can't see this objection, as consumers love to know the percentage of the grape varieties and of course don't subscribe to 'bigger numbers make better wines.'" He then went on to downplay any idea of a crisis, as "neither winery nor consumer should anticipate significant changes. And if changes are adapted, they will not be effective for years."[20] John agreed with him on the time factor and predicted that none of the proposals would take effect for at least three to five years.

De Luca, in a typical political compromise mode, then suggested that the institute could back "mandatory basic and clear information for the public at large, but voluntary use of specialized wine facts for the wine enthusiast and connoisseur." He went on to say that this "approach would permit vintners who wish to incur additional expense to offer additional information on grapes, climate, cellar treatment, and other details, pushing the cost to the more fastidious wine drinker."[21] John had weathered his first crisis, and months later he cautiously celebrated victory as the ATF dropped its labeling proposals because "of wide criticism by those in and out of the wine industry." De Luca responded, "We are pleased that the government listened to us and that they understand we are trying to do what is best for the consumer."[22] But this initial attempt to mandate the content on wine labels would soon move in a very different direction.

The alcohol labeling issue surfaced again in 1978 when Dr. Ernest Noble, director of the National Institute of Alcohol Abuse and Alcoholism (NIAAA), failed to get Congress to legislate for warning labels on alcoholic beverages because of the possibility of in vitro harm to fetuses. Noble cited medical research that suggested serious fetal damage if pregnant mothers used alcohol, drugs, caffeine, or tobacco. The bottom line was that Noble believed that there was no healthy form of moderation for pregnant women or women thinking of starting a family.

Like many things in law and politics, there are those who after defeat modify their cause in order to continue their original battle. Not willing to give up his anti-alcohol beliefs, South Carolina senator Strom Thurmond pushed an alcohol warning-label bill through the United States Senate. California senators Allan Cranston and S. I. Hayakawa had vigorously fought and failed to get a rider attached to the bill that gave an exemption for beer and wine. De Luca followed the proceedings and paraphrased Thurmond's attitude: "If you're putting a warning label on my home industry, tobacco, then there should be a warning label also on alcoholic beverages."[23] The bill failed to get enough votes in the House of Representatives, but to appease Thurmond and anti-alcohol forces, congresspeople created a bill to establish a comprehensive study on alcohol. The ATF, the Food and Drug Administration (FDA), and the Health, Education, and Welfare (today's Health and Human Services) department then tasked the NIAAA with completing the study. In June 1980 they presented their results to President Jimmy Carter and recommended that labels were not warranted.

Code of Advertising

As the labeling turmoil wound itself through the legal and governmental processes, De Luca began to address other concerns of anti-alcohol forces. The dry proponents continually argued that alcohol manufacturers were aggressively marketing "booze" in an attempt to lure people, especially children, to the evils of alcohol. To address the accusation, the Wine Institute adopted a Code of Advertising.[24] The industry had first adopted informal advertising principles in 1949, but the original codes did not meet the current societal needs for social responsibility. John believed that by updating the code, it would show detractors that the industry was serious about the responsible use of wine. In his words, "We specifically took into account the complex issues raised by consumers, those involved in alcohol abuse problems, media, educators, and government." He added that the code was "not created in reaction to any imminent government regulation" and that he wanted "the wine industry to be part of the solution and not part of the problems of today's society."[25]

In its preamble the voluntary program ensured that subscribing members wished "to focus on wine's history and tradition, its agricultural origins, the skill and artistry of its creation, its place in cooking and as a mealtime beverage, its role in social situations, and the importance of enjoying wine for sensory experience and not for the effect."[26] The code drew heavily from the groundwork laid by Leon Adams and fitted perfectly into the wheelhouse of John's ethnic, family, community, religious, and personal experiences.

The code's nine guideline laid out specific goals for socially responsible consumption of wine. It emphasized that advertising should direct viewers to the mealtime and food experience of wine, avoid any mention of excessive drinking, and promote the moderating effects of wine. Members who voluntarily abided by the advertising code also agreed to avoid using models under twenty-five, using popular music, endorsing of heroes or cartoon characters, advertising in kids' magazines or movies, promoting drinking and driving, degrading any group of people, and presenting drinking wine as a rite of passage.

The new code then emphasized the historical, religious, ethnic, and pharmaceutical properties of wine. In the code De Luca wrote that wine was an agricultural crop that came to America from European wine-drinking cultures and that modern anti-alcohol forces had degraded the mealtime and moderating attributes of a wine culture. He proposed that America was now in a position to return wine to its original status as a beverage of health and moderation and to complete the process of rebuilding a tarnished American wine culture. In a true De Luca style that was academic, political, ethnic, and media savvy, John had started his own wine-culture reeducation program to increase the sales of wine.

The code was not without its detractors. Many argued that it should not be voluntary for the institute's 262 member wineries. They also worried that corporate nonmember wineries, like Heublein, United Vintners, Beaulieu Vineyards, and foreign wineries, would not be subject to the plan. Further complicating public trust of the plan were the legions of anti-alcohol detractors who emphasized the fact that under federal law, the institute's plan had no enforcement provisions. Critics included NIAAA director Dr. Ernest Nobel, who felt the code was not strict enough.[27] Even some of the members of the Wine Institute disliked the code because of the negative tone of the document, which made it sound like a biblical injunction with a list of "shall nots." They argued that it put the industry on the defensive because it raised issues that most wine drinkers had never thought about.

De Luca responded to these concerns: "We expected some interest in the trade press, of course, but it became a media event overnight. Obviously, there is far more interest in this issue than anyone expected."[28] This whole process served as a learning experience for De Luca. He had put his finger on the pulse of what would be a major battle for the rest of his career: health, alcohol abuse, and anti-alcohol forces.

Through all the fracas, John stayed the moderation course and honed the institute's position. In his words:

> [I]t leads into a mode of thinking that I believe has become a real strength for the industry. The excise tax, warning label, all of these advertising, drunk driving issues, international affairs – there is a common theme that runs through all of them, and that is analyzing the issues, taking a position so that you can advocate a position, not creating a vacuum, and, above all, not permitting either government or adversaries to frame the issue for you. So there is much to be said, comparing what we did on the excise tax to what we did on the advertising code. That is, what is it that we believe

must be done? What is it, before somebody takes a position against us, that we should stand for? How do we frame the issue so that when we go to the public, when we go to the community, when we go to the media, when we go to the elected officials, we stand for something, rather than being against something.[29]

Declaration of Principles

Following his food and moderation line of thinking, De Luca took a proactive approach to educating the public, leaders, and agencies. In his words, "Cultural ignorance is our greatest enemy in this society. That's really been the millstone around our neck—not enough people really knowing what it was we've inherited and what it is that we're doing and contributing." To respond to the evolving playing field, De Luca created a new management style for the institute that included long-range planning, white papers, political lobbying, media engagement, increased winery involvement, and the philosophy that "you don't really win battles as much as you shift the terrain. It's a never-ending process of always being vigilant."[30] To begin this new direction, De Luca enlisted Wine Institute staffer Patricia Schneider to research and document more than four hundred groups that dealt with alcohol issues.

On June 23, 1978, De Luca released his first white paper, "Declaration of Principles," for the institute's Wine Media Day. By this time he had an opportunity to reflect upon a strategy to proactively strengthen the idea of an American wine culture. Like a political leader, he laid out the planks of his platform. In his words: "In preparing for this day, I drew up these principles as I perceived them. They can be debated and they can be modified and refined. In fact, I invite their review and critique by vintner and writer and all parts of the industry. But it seems to me that the time is right to discuss for the first time in contemporary terms a new concept that I am calling 'A Declaration of Principles.'"[31]

The document overtly reflected the principles that De Luca had learned from family, religion, politics, academics, and experiences writing papers as a White House Fellow. He laid forth ten criteria for rebuilding an American wine culture. In his declaration he portrayed wine as a natural agricultural product that had enhanced food for thousands of years. He reinforced the idea that wine is a healthful product and that the institute would continue to support research on the beneficial aspects of the beverage. The paper also highlighted the historical moderating effects of wine and vowed that the institute would be socially responsible and would work toward helping to resolve alcohol-abuse issues. De Luca also promised to assist in future industry growth by finding political solutions for both domestic and foreign trade problems in order to achieve reciprocity and equity.

Miraculously, he had kept his job, and, more important, he had laid out a plan for rebuilding the image of wine in American society. Journalist Eric Brazil observed, "California, which dominates the American wine industry like a whale in a swimming pool, is starting to cultivate a new image as a friendly giant." He quoted De Luca: "In the long run domestic protectionism won't help the wine industry, only increased consumption will."[32] At that time, twenty-nine states produced wine, and California led with the production of 90 percent of all American wine. But because of protectionist barriers left over from repeal, the California industry did not have equal access to many states. De Luca doubted that the equity issue could be solved at the federal level and saw the boosting of domestic consumption as the only alternative to state anti-alcohol protectionist barriers. During his initial crisis management, De Luca had established a distinct leadership pattern, based upon research, reading, writing, lecturing, and face-to-face encounters.

{7}

Neoprohibition

The Continuing Battle for an American Wine Culture

> To think bigger than ourselves.... You've got to concentrate on your
> ten acres, you've got to really cultivate those ten acres, but understand
> where it fits into your valley, where that valley fits into your region,
> where that region fits into our state and in our country.
>
> —John De Luca, "President and CEO of the Wine Institute," oral history

While stabilizing the Wine Institute, De Luca understood that the ele-
phant in the room was the never-ending threat of anti-alcohol adher-
ents. As concerns over fetal alcohol syndrome escalated, wine had lost
its image as a moderating beverage. This allowed prohibition defenders to intensify their attacks by asserting that all alcohol was a threat to
the health of the nation and now promoted the idea that anyone who
questioned this premise was sacrificing the health of Americans. Unlike
the pre-Prohibition drys, who wanted to end all alcohol use, the new
anti-alcohol proponents sought ways to limit the use of alcoholic beverages. They planned to accomplish this task through the use of age
restrictions, public consumption rules, driving restrictions, warning
labels, increased taxes, new regulatory laws, and anti-alcohol health
policies enforced by the federal government. Their ultimate goal was
to demonize alcohol by relabeling it as a sin, a health hazard, and an
anti-American substance.

ALCOHOLISM

To understand the new alcohol scare and De Luca's response, it is best
to contextualize how alcohol public policy evolved between the 1930s
and 1980s. As Prohibition failed and the Great Depression ravaged
the national economy, consumption of alcohol increased, and many
began to worry about alcohol abuse. One sufferer, Rowland H., had
visited the Swiss psychoanalyst Carl J. Jung for assistance in combating

his alcohol addiction. Jung referred him to Frank Bachman's Oxford Group, which used Christian principles to overcome the urge to drink. While attending meetings at the Oxford Group, Rowland met fellow alcoholic Ebby T., and the two began a quest to assist fellow addicts. Two of their mentees included Manhattan stockbroker Bill W. and physician Robert S., who in 1935 founded Alcoholics Anonymous. The founders never used their last names because of societal norms that identified all alcoholics as mentally and morally weak. An immediate part of their mission became an educational program to show others how to use their disciplined solutions to overcome their addiction to alcohol.

In 1944 Marty Mann founded the National Committee for Education on Alcoholism, which became the precursor for today's National Council on Alcoholism and Drug Dependence (NCADD). The new organization attempted to combat continued increasing rates of alcoholism prompted by the end of World War II and the rise of the prosperous Cold War economy. That same year, the United States Public Health Service recognized alcoholism as one of the most serious threats to American public health and named it as a workplace problem that threatened national security. In response, Congress passed the Alcoholic Rehabilitation Act of 1947 and began the process of redefining alcoholism as an illness. With the assistance of the new congressional act, NCADD partnered with Con Edison, DuPont, and other companies to provide employee alcoholism programs designed to make their workforce more productive and to improve employees' personal lives. As awareness increased, Lois W. increased the reach of the alcohol programs by assisting the families of alcholics with her new Al-Anon program.

A breakthrough for treatment of alcohol abuse came in 1952 when the American Medical Association defined alcoholism as a disease. Despite the new designation, past practices lingered, and many in the medical community hesitated to accept the problem as an illness. To bolster their decision, the AMA doubled down and passed a 1956 landmark resolution calling for broad acceptance of admitting alcoholics to general hospitals and urging hospital administrators to provide adequate and appropriate services. The disease theory reached critical mass in 1960, when medical researcher E. M. Jellinek published his article "The Disease Concept of Alcoholism," and in a short time the disease model became the standard for treating alcohol addiction. But the biggest move to assist alcoholics came in 1967, when the AMA passed a resolution identifying alcoholism as a "complex disease" and pushed the medical community to develop treatment procedures.

To assist in the development and expansion of treatment programs, President Richard M. Nixon signed the Comprehensive Alcohol Abuse and Alcoholism Prevention, Treatment, and Rehabilitation Act of 1970. The legislation, also known as the Hughes Act—named after the recovering alcoholic Iowa senator Harold Hughes—authorized a comprehensive, federally funded program to underwrite research and to promote prevention and treatment programs. The act also established the National Institute on Alcohol Abuse and Alcoholism (NIAAA) as a part of the National Institute of Mental Health. In 1974 NIAAA became an independent institute and part of the National Institutes of Health.

In America the acceptance of alcoholism as part of new public health awareness gained momentum, and policy makers throughout the 1970s pushed for more government action. In response, Congress utilized scientific data and assigned millions of dollars in federal funds to sponsor special research centers. The problem for many in the medical community was that these new centers based their programs on the controversial 1950s "distribution theory" formulated by French social scientist Thomas Ledermann.[1] The theory proposed a correlation between the number of heavy drinkers in a region and the total per-capita consumption in that region. Wolfgang Schmidt and Jan de Lint, from the Action Institute in Toronto, Canada, adopted Ledermann's work and promoted the idea that the only way to curb alcoholism was to restrict consumption of alcoholic beverages by increasing alcohol taxes, limiting the hours of sales, cutting the number of outlets, eliminating all advertising, and creating government-supported anti-alcohol campaigns. Needless to say, their interpretation of the distribution theory presented a huge roadblock to further growth of the wine industry and provided new fodder for anti-alcohol supporters.

As the research for causes and treatment of alcoholism progressed, new concerns about public health surfaced. In 1973 investigators published reports about fetal alcohol syndrome as a cause of birth defects in children born to alcoholic mothers. The issue raised enough public concern that it energized the NIAAA to sponsor a fetal alcohol syndrome workshop, and soon after, Dr. Ernest Noble, head of the NIAAA, began a campaign to require warning labels on all alcoholic beverages. The problem for the wine industry was that the push to warn pregnant mothers, through state and federal legislation, left no room for any form of moderation. This pushed wine further into the same category as all distilled spirits, tobacco, and drugs.

NEOPROHIBITION

Given the increasing political pressure, health concerns, and resurgence of anti-alcohol forces, De Luca decided to proactively address the negative publicity facing the wine industry. To move forward, he researched the topic, and in 1976, he released a white paper, "The New Prohibitionists: What They Say and How They Affect Legislation and Government Policy," to the media and Wine Institute membership. In the first part of his treatise, De Luca briefly laid out the history of how anti-alcohol forces had achieved the Prohibition Amendment by appealing to the morality and religious fervor of fundamental Christians. Their anti-alcohol crusade had grassroots success and heavily influenced congressional elections in the early part of the twentieth century. This revitalized religious movement allowed anti-alcohol proponents to make or break the careers of lawmakers by labeling them as pro- or anti-alcohol.

According to De Luca, this restrictive ideology lessened when the repeal era allowed a "wet" urban society to place alcohol "policymaking in the hands of bureaucrats in regulatory agencies and the federal government played no leadership role in supporting dry forces."[2] As a result, politicians in the latter half of the twentieth century relied upon specialized scientists and research centers for data to create government public health policies. Upon realizing their loss of control, a revitalized group of anti-alcohol devotees, "neoprohibitionists," regrouped and again laid responsibility for many disastrous social problems and health issues on all alcoholic beverages.

De Luca understood the disastrous effects that the revitalized anti-alcohol movement could have on the California wine industry. In the 1975 California legislative session, Senator Arlen Gregorio, who served on the Senate Health and Welfare Committee, introduced SB 204 in an attempt to raise taxes and earmark the funds for alcoholism programs. This would have been the first earmarked taxation measure in California history, and Governor Edmund G. Brown vetoed the bill because he had made a campaign promise of no new taxes. Not willing to give up, Gregorio then pushed through legislation that created a $1-million fund to support anti-alcohol advertising. Since the bill required no tax increases, Governor Brown signed Gregorio's bill and furthered the program's power by reorganizing the Office of Alcoholism and making it part of the Department of Alcoholic Beverage Control. This placed California alcohol production and sales under the purview of one agency. As the governor's plan moved forward, he named Rita Saenz, past contractor for the National Council on Alcohol, as director, and she immediately announced that the goal of her office would be to reduce the consumption of alcohol.

The angst over the new alcohol theories and policy directions grew when federal congresspeople began to make statements reflecting talking points espoused by neoprohibitionists. In one instance, Maine senator William Hathaway, chairman of the Senate Subcommittee on Alcohol and Drugs, clearly expressed the anti-alcohol talking points by stating that even moderate drinking posed a threat to public health. He then began a series of hearings on alcohol advertising. As the message spread, Senator Edward "Ted" Kennedy of Massachusetts showed an interest in Senator Strom Thurmond's bill to place warning labels on all alcoholic beverage containers. De Luca responded to the new threats by releasing the white paper "Wine and Government" at a June 23, 1978, Wine Media Day event. He strongly warned the industry, and the nation, about one-sided government policies, and he advised the media to "get to know government as you know wines."[3]

On the surface, these attempts to reduce alcoholism through reducing consumption seemed reasonable. Many in the wine industry feared the extent to which more radical anti-alcohol proponents would push reasonable boundaries. For De Luca, the real question became: "What motivates the new prohibitionists?" He believed that the new movement contained a cadre of moral and religious leaders, like those from the Prohibition era, and he worried that they would yet again harass and threaten the careers of politicians and academics who needed to protect their reputations and government grants. By marginalizing researchers, these anti-alcohol groups had built "a fail-proof system, whereby they will judge whether or not their theories are effective.... The foxes, some would say, get to watch the chickens."[4]

The neoprohibitionist movement funded groups like the American Business Men's Research Foundation, which produced periodicals and successfully lobbied many congresspeople for anti-alcohol policies and laws. Members of the renewed prohibition forces quickly learned that in order to control the conversation, they simply needed to control the funds of alcohol professionals—counselors, physicians, and program administrators. These professionals needed grant funding for their programs, and they bowed to those who controlled the congressional purse strings.

Embedded in De Luca's moderation beliefs was a message of hope for the wine industry. He believed that the most influential and respected private and volunteer organizations did not support the new prohibition theories and political policies. As proof, he offered up the fact that the National Council on Alcoholism supported third-party research on the affects of alcohol consumed in moderation. They had taken a supportive approach to alcoholism and through their congressional lobby had pushed for policies and legislation to help individuals receive assistance for alcohol-related diseases. The organization believed that alcohol abuse was an individual's problem and not the societal problem, articulated by the neoprohibitionists. Further support came from the Alcohol and Drug Problems Association, a trade organization for alcoholism treatment centers, which opposed earmarked taxation and preferred general government funding that fostered research on all sides of the issue. To show their support, the Education Commission of the States presented a paper to President Jimmy Carter that recommended "education and community-coordinated information, and rejected the new prohibitionist theory and control measures."[5] Dr. Morris Chafetz of the newly founded Health Education Foundation opposed all the neoprohibition programs and believed that moderation presented the best alternative. De Luca added to the list of supporters groups like the North Conway Institute of Boston and the U. S. Jaycees, which backed the moderation cause.

For De Luca, the immediate concern was how to get third-party government-funded research to make the case for moderation. This would prove to be difficult, since Dr. Noble of NIAAA had cut all funds for research on moderate drinking. John observed, "This means that the groups seeking to build a national consensus around alternatives to the new prohibition theory are increasingly turning to the alcoholic

beverage industries for support and funding."[6] This presented a problem for De Luca, who preferred independent research because industry-
funded projects would not hold the same persuasive weight.

TAKING ACTION

As the battle escalated, De Luca drew strategies from his university,
ethnic, political, religious, and family values to design a course of
action aimed at rebuilding societal beliefs in moderation and the idea
of an American wine culture. To get his message across, John began an
active campaign aimed at the halls of government and wine drinkers.

In a 1976 speech to the House of Representatives at Working Group
on Prevention, Education, Information, and Training of the Interagency
Committee on Federal Activities for Alcohol Abuse and Alcoholism,
De Luca addressed wine advertising and health issues. He assured the
committee that the wine industry wished to advertise only "to adults
who already are consuming alcoholic beverages" and reminded them
that in 1976 the wine industry spent only $64.3 million on advertising
compared to the tobacco industry's $279.3 million, breweries' $133 million, and $213 million spent by distilleries.[7] He presented statistics pre-
sented to the committee showing that over the five-year period between
1972 and 1976, the average American adult consumption rate had only
gone from two gallons to two and one half gallons per year. He also
assured the committee that the wine industry's intent was to promote
moderation and the role of wine as part of a meal.

De Luca kept on message and repeated his mantra of drinking in
moderation and rebuilding an American wine culture whenever and
wherever he could. At a 1977 address to the 117th quarterly dinner
meeting of the Society of Medical Friends of Wine, De Luca delivered
a talk titled "The Progress of Wine in America."[8] He emphasized the
need to continue messaging about the moderation, health, and establishment of an American wine culture. In another effort to bring the
medical community to his side, De Luca presented the white paper
"Wine Health and Society" to a University of California Medical Association symposium. The meeting brought together University of California medical schools, the Society of Medical Friends of Wine, and
doctors and vintners including David Bruce, Joseph Carey, Klaus Deh-
linger, John Staten, Martin Griffen Jr., Stanley Hoffman, William Casey,
Bernard Deps, and Austin Steen. The message of wine as a social and

moderating part of everyday meals carried the day, and most agreed that America seemed to be slowly rebuilding its lost wine culture.[9]

On May 3, 1980, a repeat drunk driver, who had just been released from his fourth DUI arrest, struck and killed Cari Lightner, a Fair Oaks, California, teenager. In her sorrow Candance Lightner, Cari's mother, set out on a crusade to protect other children from untimely deaths by drunk drivers. To get her message out, Lightner created the national organization Mothers Against Drunk Driving (MADD) in October 1980. As a parent, John felt the mother's pain and personally deplored the excessive drinking that caused the accident. De Luca believed, "There is no greater tragedy on the highways than this sense of alcohol abuse. I've always advocated that we should be at the forefront of that development. Therefore, my wife and I became members of Mothers Against Drunk Driving. We've worked very well with the organization, and we have seen no contradiction between our association with the wine industry and being responsible parents and corporate citizens."[10] To prove his personal point, De Luca encouraged Wine Institute members to join MADD and had the institute provide grants for MADD's educational programs. He saw the organization as a reasonable group of people who, unlike the neoprohibitionists, did not want to stop all alcohol use. He recalled:

> The neoprohibitionists genuinely want to attack the product in terms of the integrity of the product. There's a very important distinction. I think we all can make common cause with Mothers Against Drunk Driving and any other group against alcohol abuse. The insidious part of the neoprohibitionist agenda is to say there is no such thing as responsible drinking, there is no such thing as moderate drinking; the product itself is inherently dangerous. Therefore, any consumption is dangerous consumption. Of course, that is refuted by history, that is refuted by common sense, that is refuted by civilization. So there are organizations that we can work with on the question of alcohol abuse, as against this politically rigid approach of trying to say there is no responsible behavior, or that there is no distinction between, say, wine, beer, and spirits and crack cocaine. That's really what is happening—That's the tug of war that's occurring.[11]

Despite the active measures pushed by anti-alcohol forces, the production and consumption of American wines increased over the next few years, and in 1981 De Luca released a white paper titled "Fine Wine and True Grit." His paper outlined the success of the institute's moderation campaign and of attempts to return wine to its rightful place in agriculture. Exports of California wines had jumped from 1,398,000 gallons in 1976 to a 1980 high of 7,905,000 gallons. Not only were Americans drinking more better-quality premium California wines, but so were wine-drinking countries around the world. John explained the growth this way: "At this moment the United States is experiencing lifestyle changes favorable to wine. Men and women interested in nutrition, moderation, cooking, travel, self-help and improvement, all forms of active sports, are merging with what were previously more ethnic styles of enjoyment and a significant acculturation process. Contemporaneously, people in all stations of life, in different parts of the globe, opinion makers and decision makers, are saying quite a bit more than nice things about California wine."[12]

De Luca's premise was sound if one places it in the context of the explosive growth of the California-cuisine movement that grew from the Bay Area and in a short time encompassed the United States and eventually the world. Americans were looking for fresh, seasonal, local foods and compatible wines for their California fusions of French, Italian, Spanish, Mexican, and Asian recipes. Chefs, restaurants, and home cooks throughout America began to pair all their new dishes with wine.[13]

But all of the news was not good, and the anti-alcohol forces continued their fight against the perceived evils of alcohol. In 1980 the federal government, at the behest of the Center for Science in the Public Interest (CSPI), ended a decade of indecision and ruled that by the year 1983, all beer, wine, and liquor would have ingredient labels. De Luca responded to the new ruling by commenting that "the regulation has rough edges, the most onerous of which is the fact that apparently the government cannot assure that the regulation will be enforced overseas. . . . We abhor this double standard." He then declared that the industry would work to overturn the ruling.[14] Within the first year of the alcohol-friendly Ronald Reagan administration, government officials dropped the label requirement before it had a chance to take hold.[15]

Anti-alcohol believers were devastated by the Reagan decision, and in a January 4, 1982, *Washington Post* opinion piece, economist Peter Navarro reopened the argument for wine-ingredient labeling. Navarro argued that the Reagan administration's love affair with laissez-faire capitalism had killed the labeling of wine, much to the detriment of the consumer. Navarro questioned the Reagan administration's policies and asked how consumers could make wise decisions about quality, health issues, or value related to cost without proper labels. In his words, "As economists have shown, the free market is best if and only if it is perfectly competitive and both buyers and sellers have complete information."[16] This attempt to resurrect what De Luca had presumed was a past issue infuriated the industry and drove John to write a 1982 white paper titled "Setting the Record Straight on Wine." John blasted the idea of labeling wine and declared:

As wine growers, attuned to the rhythms of nature and the harvest, California vintners are alert to the ravages of frost and flood, heat, and pestilence. Natural enemies are an unfortunate reality in the world of agriculture. However, they also provide good training for the afflictions of activist politics, For example, it appears from recent articles, orchestrated by the Center for Science in the Public interest (CSPI), that we wine people have been targeted for a new infestation. The biting edge of the attack on our integrity is the irresponsible distortion of the "ingredient labeling" issue and the misrepresentation of the Administration's decision could descend into a costly and unwarranted regulatory burden.[17]

De Luca continued his forceful argument by accusing Navarro of becoming a "convert to a zealous cause" espoused by people who act as though "they have a special dispensation from the requirements of truth."[18] De Luca was correct in insisting that no California wine contained dyes, oxblood, urea, foreign substances, or suspected carcinogens, as Navarro had also alleged. A few years later ATF regulators found diethylene glycol (used in antifreeze) in wines from Australia, Austria, West Germany, and Italy, but not in wines from the United States.[19] American wineries followed federal regulations that defined table-wine as follows: "Grape wine is wine produced by the normal alcoholic fermentation of the juice of sound, ripe grapes."[20] John finished the white paper with the Latin phrase *in vino veritas*—in wine there is truth.

American wineries took pride in the quality of the wines they were producing and believed they had achieved equity with European Old World wines. To prove their point, the Wine Institute sponsored a Paris tasting a few months after the Navarro incident. Fifty-two American wineries poured wine at the American embassy in Paris for more than two hundred French restaurateurs, wine makers, and the press. Those tasting at the event commented on the quality of the American wines, and the embassy's Foreign Agricultural Service spokesperson remarked that the "enological event was intended to gain favorable exposure for another worthy American product in an important foreign commerce market."[21] In the midst of the distractions created by the anti-alcohol forces, American wine standards were making an international splash. To further raise the recognition for quality domestic wines, the Wine Institute sponsored numerous domestic wine tastings in what they referred to as "Celebrating the marriage of food and wine."[22]

More startling news came in 1982 when NCADD called for increased alcohol taxation, a national minimum drinking age of twenty-one, more regulated alcohol purchase laws, and health-warning labels on all alcoholic products—goals that were all achieved by 1990. Their anti-alcohol message gained an A-list supporter when former first lady Betty Ford lent her name to a treatment center for alcoholism and other drug addictions. The new anti-alcohol successes hit home, and De Luca again realized that while prohibition may have been defeated, the anti-alcohol forces were far from giving up the fight.

But De Luca had another worry on the horizon. Alcohol corporate giant Joseph E. Seagram & Sons had purchased the majority of Coca-Cola's wine properties for $230 million, making them the second-largest Wine Institute member behind E&J Gallo.[23] People inside and outside of the industry worried that large corporate ownership of wineries would degrade the message of drinking wine in moderation. History had taught winemakers that vintibusinesses valued profits over messages of moderation and feared that a pure business model would tarnish wine's image, much as it had before Prohibition.

For the industry, it was hard to make a case against the deal, since California wineries had just finished a rain-soaked season, faced flat retail sales, and endured skyrocketing land prices, and many needed additional money to modernize production facilities. The historic, cyclical pattern of rebuilding with an influx of corporate money to invigorate sales and update infrastructure had worked in the past, and many felt it could work in the future.[24] Of most concern to the Wine Institute was the fact that anti-alcohol enthusiasts took advantage of the wine industry's increased corporatism to reinforce their anti-alcohol attacks. In a short time, De Luca's corporate worries proved correct.

In a move that aided the anti-alcohol forces, Seagram executives ran an advertising campaign based on the equivalency idea that "a drink is a drink." They promoted the idea that twelve ounces of beer, five ounces of wine, or one and one-quarter ounces of whiskey all had the same effect on consumers. This campaign angered many in the industry, and the Winegrowers of California responded with: "The Seagram advertising campaign ignores the fact that wine, beer and liquor produce significantly different rates of absorption, blood alcohol levels and resulting intoxication when abused."[25] Leon Adams agreed and called the equivalency claim "a great big lie."[26]

At first De Luca followed his past practice with corporate wineries and remained quiet about the issue. As the tension within the industry increased, he weighed in by saying: "Seagram's policies were in conflict with the fundamental tenets of our organization." He continued: "Equivalency was a ploy to persuade congress to levy the same taxes on beer and wine that it did on spirits." De Luca revealed that Congress had just passed an excise tax on distilled spirits that would increase prices by about fifty cents per bottle.[27] Seagram executives believed that if they could make the case for equivalency between distilled spirits and beer and wine, then it would make sense for those industries would also face the same increase. The rift intensified, and Seagram responded by withdrawing from the Wine Institute.[28] The dispute also played into the hands of anti-alcohol proponents.

Most wineries supported De Luca's moves toward building an American wine culture. The Winegrowers of California held seminars in 1985 and 1987 to discuss issues surrounding the neoprohibitionist agenda, and during one meeting Robert Mondavi announced his mission to create a coalition to defend wine as part of a civilized lifestyle. Mondavi then turned the idea over to the National Wine Coalition, a subgroup of the Wine Institute, to complete the task. Because of a lack of funding, the coalition never succeeded in establishing itself as a national organization and subsequently dissolved in 1995. The American Wine Alliance for Research and Education also dedicated itself to an educational program to create a "balanced, comprehensive view of wine in the United States." Regretfully, it also stalled.[29]

If the wine industry wanted to scale up, it was obvious that they needed to increase sales through expanded consumption. The one giant domestic roadblock was the Repeal Amendment that allowed every state to determine how to distribute—or refuse to distribute—alcoholic beverages within their own boundaries. This left America with fifty sets of complex distribution requirements that restricted California wine sales. In response, De Luca initiated a campaign to find legal ways to circumvent these various state laws and help wineries increase out-of-state sales. In one attempt, he proposed that California wineries sell bottled wine directly to private Iowa liquor businesses and bypass the state-mandated stores. Utilizing data from a Wine Institute study, De Luca predicted that sales would increase by more than $12 million per year and assured Iowans that the "escalation in wine drinking would not increase alcoholism in the state." Rolland Gallagher, director of the Iowa Beer and Liquor Control Department, pushed back at De Luca and declared, "Wine sales very well may triple under private enterprise, but the state would not gain revenue, alcohol abuse would increase, and retail prices would be higher than they are now."[30] De Luca doubled down on his prediction and retorted that with a modest excise tax of 70 cents per gallon, the state could bring in much-needed money and create 1,120 new jobs. Needless to say, the heated discussion resulted in no actions, and John learned that the idea of free wine trade in all fifty states would involve a steep uphill battle.

Advertising of beer and wine came under further scrutiny in the mid-1980s. A petition with more than six hundred thousand signatures called for the United States Senate to ban all alcohol advertising from radio and television. Stephen Chapman of the *Orlando Sentinel* described the petition by Project SMART (Stop Marketing Alcohol on Radio and Television) as the "work of a coalition of abstainers, self-styled guardians of the 'public interest and compulsive busybodies.' Supporters of the petition included the Mormon and Baptist Churches, the Center for Science in the Public Interest, and the National Parent-Teacher Association."[31]

De Luca viewed the kerfuffle as an opportunity to tout the Wine Institute's Advertising Code of Ethics. His response came on February 7, 1985, as he addressed Florida senator Paula Hawkins's Senate Subcommittee on Alcoholism and Drug Abuse. John told the committee about the Wine Institute's advertising policies and stayed true to his continuous talking point about rebuilding an American wine culture. He also told the committee that the wine industry wished to "rejuvenate our recent history where wine has been the principal beneficiary of the cultural movement in America emphasizing nutrition and self improvement and the dynamism of women and the social emphasis on beverages with less alcohol."[32] He reiterated the Wine Institute's Advertising Code of Ethics and reassured the committee that the institute wished to actively assist in the process of helping alcoholics to recover. De Luca emphasized that the code promoted the moderating role of a mature lifestyle centered on meals and celebrations. John also warned Hawkins, and the committee, that there were those who pushed a political agenda to battle against alcohol in a polarizing way. He finished his testimony with a promise of continued support for alcohol-abuse programs: "In order for us in the California wine industry to be perceived as part of the solution and not part of the problem we must do credible things. I submit that our Code and its implementation, our position on tougher law enforcement and drunk driving, the twenty-one year maximum age for public purchase and possession, our initiatives as represented in our exhibits, have been credible commitments to the American public."

As the anti-alcohol campaigns increased, many across the nation began to take an interest in De Luca's arguments. Augustus H. Hewlett, president of the Alcohol Policy Council (a liquor industry–financed group), warned that "the liquor industry is threatened by a dangerous national backlash" that he described as neoprohibitionism.[33] *Cincinnati Enquirer* journalist Thomas Gephardt agreed in his article "Here Comes Neo-Prohibition" and lamented, "Yet, our grim experience with Prohibition and our changing attitudes about alcohol haven't dampened the zeal of a new crusade by Neo-prohibitionists to change the way the liquor business runs." He argued that if one followed recent government actions, it was obvious that anti-alcohol forces were at work. To prove his point, he listed their current objectives as raising the drinking age to twenty-one, requiring health warning labels, imposing additional taxes, banning beer and wine advertising on television, making tavern owners responsible for drunks, supporting MADD, and restricting bar and restaurant happy hours. Florida newspaperman Andrew Glass also agreed and wrote, "The new Prohibitionists tend to be middle-class types who seek to create an atmosphere of social disapproval for all but the mildest of drinkers. If they had their way, the national concoction would be a 'spritzer'—a splash of white wine diluted by plentiful dollops of club soda."[34] John's early attempts to seek media support had begun to pay off.

De Luca's daily task became one of reciting his themes of moderation, wine as food, and the need for an American wine culture. He stayed on message and determined to stay in the fight for the long haul through his white papers to institute members, appearances in the halls of Congress, and releases to radio, television, and print media. He had learned the lesson that this was an ongoing battle and that he must stay on task and not let his guard down. In his words:

> But what we really want you to do is not micro-manage an office but to think through the big issues, the strategic issues, these commanding heights issues. Over a period of years, through board of directors discussions and back and forth with officers and with members, this line of discussion was followed: we will fight many skirmishes, we will put on thousands of wine tastings, we will release a lot of energy and bring tourists to our vineyards, and that will be beneficial in and of itself. But we have to break out of the enclosure, and the definition that's been imposed on us, the political, social, and even media observation of us was either through the "booze," through the term "liquor," "sin tax," "gateway drug," "drug of choice.". . . Unless we break through this entrapment that we're in, which is both terminology and ideological and highly emotional, we're going to have this constant pounding that we're a carcinogen, a reproductive toxicant, and if successful in our strategy then the focus will be exclusively on the abuse of the product. We have to redefine ourselves.[35]

{ 8 }

Alla Vostra Salute—To Your Health

> We've been operating in terms of some very specific social issues, most importantly the question of alcohol abuse, alcoholism, social responsibility on the part of citizens, social responsibility on the part of the industry, and moderation. We see interrelated not only the question of alcohol abuse and alcoholism, but the whole question of health, health care, and how to address issues that are legitimate, namely questions of fetal alcohol, reproductive toxicity, questions related to carcinogens.
>
> —John De Luca, "President and CEO of the Wine Institute," oral history

It became apparent that determined anti-alcohol forces would not surrender in their battle to destroy the idea of any return to an American wine culture. Understanding this, De Luca prepared the Wine Institute for a long, complex campaign that shifted their energy and talking points to be more inclusive of wine's health benefits. His concerns were real, and within a short time the next anti-alcohol battles simmered in state governments and the halls of Congress over health labels, direct sales, and the War on Drugs. Needless to say, members of the wine community saw these neoprohibitionist attacks as a roadblock that could hinder or destroy much of their industry.

THE WAR ON ALCOHOL CONTINUES—LABELS

Neoprohibitionists continued to identify wine as a part of what they perceived to be the nation's alcohol problem, and in 1988 they lobbied Congress and won passage of the nation's first alcohol labeling requirements. Introduced in the House of Representatives as HR 5210, the Alcoholic Beverage Labeling Act required that the labels on alcoholic beverages carry a government warning for pregnant women and people who operate vehicles. In 2000 dry forces then proceeded to push their anti-alcohol campaign further by convincing West Virginia Democratic senator Robert Byrd to sponsor a new bill to enlarge the size of warning labels and to include the words "May Cause Health Problems." Their legislation asked for a wider label, with large red or black letters on a white background and an exclamation point inside of a triangle.[1] The bill died in the Senate, but neoprohibitionists decided to continue the battle. In 2007 they championed a new bill to allow the Alcohol and Tobacco Tax and Trade Bureau to require all alcoholic beverages to carry a nutrition label. As part of the new proposal, these labels would follow the same mandates placed on food and include calorie counts and percentages of carbohydrates, fats, sugars, proteins, and artificial ingredients.[2] The new labeling pressure eased when the proposal failed to gain traction in Congress.

THE WAR ON ALCOHOL CONTINUES—DIRECT SALES

Undaunted by their recent federal labeling setbacks, neoprohibitionists expanded their attacks by supporting members of state legislatures who were open to the prohibition of direct-delivery purchases of California wines. Their new tactic had some success, as twenty-six states made direct consumer purchases of California wine a crime punishable with fines and jail time. This led to a rebellion by some consumers, who argued that this measure violated the idea of free trade as promised under the Commerce Clause of the United States Constitution. The model of "Bootleggers and Baptists" surfaced again, however, as alcohol wholesalers fought to protect their post-prohibition, three-tier distribution system by supporting neoprohibitionists who sought to restrict direct sales of alcohol. To this end, large distributors aggressively pressured state governments to protect their guaranteed three-tier profits by keeping wineries from transacting direct wine sales with consumers. They justified their stance by arguing that only they could ensure collection of all taxes and could keep alcohol out of the hands of minors. Fervent anti-alcohol supporters backed distributors in the hope that restricted sales would also lower alcohol consumption.

Wholesalers argued that the Twenty-First Amendment (Repeal of Prohibition) had given states broad rights to regulate the sale, distribution, and importation of alcoholic beverages. But this interpretation had led to disorder in the wine market, as California wineries faced fifty varied sets of state alcohol laws, an untold number of county laws, and no consistent federal policies and regulations. Making the matter more complicated was the introduction of fax, telephone, and internet sales that permitted consumers to purchase wines across state lines. For small wineries, these direct sales were a matter of life and death, as wholesalers consolidated and small wineries lost access to the distribution system. In a move to consolidate power, Juanita Duggan, CEO of Wine and Spirits Wholesalers of America, vowed that her organization would vigorously oppose direct sales.[3] So the wine industry again turned to the Wine Institute for assistance.

WINE MODERATION IDEAS FALTER

By the later half of the 1980s, anti-alcohol lobbyists paid for and publicly distributed questionable research as a means to force congresspeople to support laws and policies that attacked the very core of the idea of moderation.

In 1986 Los Angeles Democratic mayor Tom Bradley faced Republican incumbent George Deukmejian in a heavily contested California governor's race. Bradley hoped to secure liberal-leaning voters by adopting ideas from the environmental movement. Under the advice of political strategist and state assemblyman Tom Hayden, and his high-profile wife, Jane Fonda, Mayor Bradley supported Proposition 65, which called for placing labels on products with toxic chemicals. Bradley lost the election, but California voters overwhelmingly passed Proposition 65, and most wineries worried about the implementation of the law and its ramifications for their industry.

Proposition 65 quickly proved to be a controversial law. The architects of the bill had originally hoped that the law would protect drinking-water sources from toxic substances that caused cancer and birth defects and believed that the law would coerce companies into replacing toxic chemicals with safe ones rather than bearing a "scarlet letter" on their product. The act required that the governor draw up a list of all chemicals "known to the state" to cause cancer or to result in reproductive harm and then to ensure that consumers were warned if a product contained any of the identified chemicals. Of concern for manufacturers was the fact that enforcement fell upon the state attorney general, local district attorneys, and "any person in the public interest." In effect, this deputized citizens and allowed them to sue anyone who, they believed, was exposing Californians to toxic chemicals. What worried the business community the most was the fact that the burden of proof rested with the defendant, and no standards were established for amounts of safe exposure. Over time, plaintiff-supported science became the only way to fight toxicology zero-effect data and arrive at acceptable safe-level ingestion or exposure recommendations.[4] Agricultural products were not immune from prosecution, as can be seen by cases involving nineteen companies, including Starbucks, Bigelow, and Hain Celestial, over lead levels found in the soils used to grow tea plants.

Because wine contains alcohol and ethyl carbamate (urethane), the governor's office placed wine on the list of substances requiring warning labels. Alcoholic beverage manufacturers panicked once they realized that their products would now require warning labels that read, "Warning: Drinking distilled spirits, beers, coolers, wine and other alcoholic beverages increases cancer risk, and, during pregnancy, can cause birth defects."[5] At first no one knew what effects these labels would have on the wine industry, but De Luca believed that this application of law was certainly evidence of the growing tendency to aim prohibition efforts toward public health policies.

The processes used to determine dangerous substances under Proposition 65 became controversial when researchers, sponsored by the Wine Institute, warned that the law had flaws because it did not take into account the amount of chemical exposure or ingestion that was safe for consumers. Since the Wine Institute had sponsored the research, the findings came under fire as corporate-funded science. To counter this criticism, De Luca pushed to have third-party research with no Wine Institute ties because "we wanted to make sure that no one would claim that there was a whitewashing."[6]

Research by outside third parties proved to be somewhat detrimental for the wine industry, however. An initial study by Professor Cornelius S. Ough of the University of California, Davis, enology and viticulture program, Dr. Vincent Marinkovich at the University of California, Davis, and La Jolla physician Dr. Ronald Simond strongly suggested that alcohol use could have negative effects. Ough's team found that excessive amounts of ethyl alcohol could cause reproductive damage. In response, De Luca adapted to the situation: "When we saw that it might be a problem, nowhere near the health hazard issues that neoprohibitionists were claiming but that it might be of value, we went for an informational label."[7] In other words, he was on board with warning pregnant women of the dangers. This fallback position was not illogical, because in 1985 the FDA had declared that sulfites were a health hazard, and in 1986 the ATF had ruled that wine labels must state that the wine contains sulfites, a decision that had not hurt the industry. After decades of hit-and-miss labeling attempts, anti-alcohol proponents had won a major labeling battle.

WAR ON DRUGS AND MODERATION

Yet another neoprohibition battlefront opened when President Ronald Reagan declared the War on Drugs. Emboldened by this new policy, Senator Strom Thurmond swung into action. In his long career Thurmond had mastered the strategy of inserting small issues into larger issue bills, and he recognized a new opportunity to revisit the idea of alcohol warning labels. Thurmond's first move was to threaten to hold up approval of executive branch appointments if he did not get his label amendments passed as part of the 1987 omnibus drug bill. His riders to the bill required the use of five rotational warning labels, individual state warnings, and a clause that prohibited the alcohol industry from using the label as a legal defense in any suit arising from the warnings. Many pro-alcohol congresspeople supported the popular drug legislation and turned a blind eye to Thurmond's amendments in order to get the much-needed omnibus bill passed and signed by the president.

President Reagan signed the bill into law, and over the next two years the new law became a way for anti-alcohol defenders to bludgeon the alcohol industry. In a short time, government nutritional advisory committees proclaimed that they could not endorse drinking in moderation because they could not see any consumption of alcoholic beverages as part of a healthy diet. Wine had again lost its status as a drink of moderation, had been proclaimed to be a health hazard, and, worst of all, had been equated with drugs. For the moment, anti-alcohol forces controlled official government policy.

As a result of the new legislation, Seattle plaintiffs brought a class-action lawsuit against bourbon producer Jim Beam. This news alarmed the wine industry. In John's words:

> What if the attorneys call any of us to be on the stand? What can there be in the way of discovery? A great part of legal jurisprudence

is to throw out a net and see what you can find. So we did what any prudent organization would do. We were reassessing for ourselves, in the light of these new developments, what we put out, what we stated. It was considered not a permanent development as much as an assessment on the eve of this litigation and actually during the litigation. This period was misunderstood as a total abnegation of the past for us and that we at the Wine Institute were taking a different course of action when it came to research. We still want to find out before anybody else what there is going on in the world of the academic, medical, and lab experiments.[8]

Critics now accused De Luca of unjustified and illegitimate claims about wine's role in health and nutrition. Nancy Olsen, assistant to Iowa senator Harold Hughes, scoffed at De Luca because she claimed that his story of wine was a false narrative that he had drummed up to exempt wine from food labeling.[9] The accusations hit De Luca hard. He had always believed in his heart and soul in the truth and value of an American wine culture based upon centuries-old European traditions and experiences. This was the same man who, in 1977, had told the Society for Medical Friends of Wine that he foresaw a day when there would be wine at nearly every dinner table as a mealtime beverage to aid in digestion and in the absorption of vital nutrients. He also had faith in the idea that wine is a civilized beverage that helps distinguish human dining from animal feeding.

De Luca believed that this on-again, off-again relationship between the wine industry and government had confused the American public. he reflected:

I had officials at the FDA come to me and say, "You know, we're going to not treat you like a winery, but we're going to treat you like a pharmaceutical company, and you're going to have to go through all the hoops that pharmaceutical companies have to go through, and we're going to emphasize the drug part of the Food, Drug, and Cosmetic Act." The Federal Trade Commission came to me and said, "You know, anything that says that there might be health effects that are beneficial is misleading advertising. Because there are some people who cannot in any way, use your product, even in moderation because they're taking medications." Then the people over at the Treasury would say, "But we have legislation that covers you going back to Prohibition, and so we still have to treat you in terms of labeling, and we have to treat you in terms of taxes." So I found this whole range of different regulatory agencies reflecting this unusual history in the United States, that we were first prohibited, then repealed, then we had the states, but then we were looked upon as having to generate taxes to sustain society, and then this whole passage to, Are you food? Are you a drug? Are you pharmacologically a drug or a street drug? And what I've come to conclude is all the contradictions in this society emerged from the culture of this society, that these questions would not be raised in Italy or France or Greece the same way.[10]

Warning labels, and the push to equate wine with illicit drugs angered the majority of the wine industry, and as a result many demanded a declaration of war on anti-alcohol forces. De Luca told the *San Francisco Chronicle:* "Our history shows that these are issues we have been addressing, but it takes a collective will and the wine industry didn't have a collective will until Congress passed a bill last November requiring warning labels on all bottles of alcohol within a year. Americans traditionally 'don't respond until there is a concrete issue. It took Pearl Harbor to get us into World War II.'"[11]

RESETTING THE PLAYING FIELD

To address the new attacks, De Luca drew upon his political expertise and friendships to begin the process of shifting the policy directions that neoprohibitionists had set for the nation. John mused:

I have to tip my hat to my parents and to their ancestry, and to the Sicilian-Italian heritage that I was raised in the Lower East Side of New York. None of this is from formal education. None of this is something that I had to sit down and ponder. It came naturally, and I understood that with each year that went by, more and more there was verification of what I was arguing. It wasn't just abstractions anymore. As more and more of that was coming to pass, my little stock was rising in terms of saying, "Well you know, maybe what John told us ten years ago, maybe there was some resonance to that." So I was gaining credibility unfortunately, as things were getting dire.[12]

De Luca was not the only one espousing the health benefits of ethnic cuisines and of wine as a mealtime beverage. Just after World War II, many doctors and nutritionists had begun to discuss the concept of the health benefits of the Mediterranean diet. After observing that southern Italy had the world's highest concentration of centenarians, Ancel Keys, American biologist, hypothesized that a Mediterranean diet, low in animal fat, protected against heart disease. The results of what later became known as the Seven Countries Study appeared to show that the serum cholesterol in animal fat was strongly related to coronary heart disease.

Because of the new heart-disease studies, the American Heart Association appeared on national television to warn the American public about high-cholesterol diets. Overnight, medical professionals recommended that people adopt a low-fat diet, and this pressured government agencies to take action. In a 1986 address to a symposium on Wine, Health, and Society, De Luca addressed the topic with his talk "Health and Safety Index," which touted the Wine Institute's efforts to address health issues, advertising concerns, and alcoholism programs.[13] He stayed on point and promoted the role of wine in a healthy diet.

Adding to the wine as a healthy part of one's diet momentum was a 1986 Reader's Digest article, "The Ancient Secrets of Modern Nutrition." In the article Carol and Malcolm McConnell, a husband-and-wife team, discussed the diets of the Mediterranean people of Cyprus, southern Italy, and Greece and their health benefits that included wine as part of the regimen. After utilizing the Wine Institute's research library, they published their 1987 book, *The Mediterranean Diet: Wine, Pasta, Olive Oil, and a Long, Healthy Life.*[14] At about the same time, Patricia Schneider, former Wine Institute staff member, became the executive director of the American Wine Alliance for Research and Education and tasked her organization with fighting neoprohibitionist attacks on wine's role in healthy foodways.[15]

De Luca realized that the Wine Institute needed a means to finance a transparent strategy to elevate the narrative of wine as a moderating force and healthy part of a well-balanced diet. To accomplish this goal, in 1987 he proposed restructuring the Wine Institute and reinstituting the wine commission as a source of funds to support marketing and policy programs. The goal of establishing a new commission failed, but members supported the idea of restructuring the institute as a way to avoid having to rehash past differences and as a means of supporting the call for an American wine culture.

De Luca began the reorganization by sending out questionnaires and facilitating a series of regional meetings. He then met with individual members to build a consensus on how to move in new directions. The Wine Institute board selected Jerry Lohr, vintner and owner of J. Lohr Vineyards and Wines, to oversee a strategic planning task force, and members agreed that the Wine Institute would continue as a private voluntary organization. With that accomplished, they then rejuvenated the Wine Institute bylaws and abolished the board of directors and executive committee. In place of having 125 board members, they downsized to selecting 20 members through a system of at-large nominations and selecting another 20 by districts. Annual membership dues changed from being correlated with the price of grapes at a time of the crush to being based on a new formula where wineries could choose to pay either three-tenths of 1 percent of their gross receipts of the previous calendar year or a graduated scale of so many pennies per gallon. Under the new plan, De Luca became both the CEO and president. John recalled: "I've said, a new Wine Institute was created and the only thing that was retained from the past was the name, 'Wine Institute.'"[16] After the reorganization, De Luca made his moderation and American wine culture quest an official mission statement for the organization.

Typical to De Luca's working style, he penned a white paper titled "Wine Institute Ten Action Points." His action plan and goals now belonged to the entire institute, and the majority of members responded very well to the goals and objectives, to the discipline needed to continue, and to hard target dates with specific missions, goals, and objectives for the organization. Creation of an American wine culture now took a front seat, and De Luca began the battle to counter wine's image as a public health issue.[17]

With battle lines drawn, the anti-alcohol ranks intensified their barrage of criticism against wine as an unhealthy drug. This did not deter De Luca, the consummate political warrior, as he moved to keep the membership unified and on point. In his words:

We had history, we had tradition, but in this day's society, it wouldn't be enough to just quote what Pasteur said. He said, "Wine was the most hygienic of all beverages." Or to go back to the ancients, or to go to what I call the ancient wisdom of modern nutrition. We had to understand that the scientific community, the medical community, those with credibility, those who were not beholden in any way financially to us, they had to refute the argument, and therefore we had to call upon them through the public policy approach. Taxing us is a policy approach. To answer that, we needed to have a refutation of it. So that's why we got into the whole question of wine and health, and wine and nutrition, wine and anti-oxidants, alcohol and anti-oxidants. And so it was never a marketing program, as we were successful in encouraging third parties to come up with research, and that research was being reported by the press at large.[18]

SELF-IMPOSED TAXES

One method to take the wind out of the sail of a political opponent is to agree with them and then to propose your own program. Over the years, California industry groups had spent millions of dollars lobbying members of the state legislature to oppose bills that would increase alcohol taxes or further regulate their industry. Imagine the surprise of many in Sacramento when the Wine Institute proposed an increase in the excise tax on California wine.[19]

Anti-alcohol forces got Proposition 134 qualified for the 1990 California ballot. If it had passed, the measure would have substantially raised the taxes on alcohol, and many nicknamed the proposal the "Nickel-a-Drink Tax." Initial polling showed the measure winning by an 83-to-17-percent margin, and wineries across the state braced for the possibility of a steep tax increase. De Luca's experiences with elections and political processes propelled him to quickly put together an anti–Prop 134 coalition that included beer, wine, and distilled spirits consumers, growers, retailers, and wholesalers. But De Luca knew that it would take more than the usual allies to defeat the proposition, and he created what he called a "strategy of alliances" to battle the anti-alcohol initiative. His first move was to request that state senator Alfred Alquist (D-San Jose) and state assemblyman Dominic Cortese (D-San Jose) sponsor Proposition 12, a lower tax-rate proposal that would appeal to a large bloc of voters. In introducing this proposal, Alquist told the press, "There is a growing neo-prohibitionist movement in this state and...they hope to achieve their purpose by outrageous increases in alcohol taxes."[20]

De Luca had worked with the educational community as deputy mayor and felt comfortable approaching the California Teachers Association for assistance. To acquire their support, he promised educational leaders that money collected from his tax proposal would go to support drug- and alcohol-abuse educational programs in K–12 schools and universities. Together, they created Tax Payers for Common Sense, to fashion a coalition of the traditional supportive wine groups, minority groups, teachers and the educational community, and law enforcement organizations. John described this political move as "one of the model moments in American history when various forces are coming together and working together and needing to work together to defeat a common threat. So, I would say that it is an example that we should not forget when the question comes up about distinguishing ourselves." Proposition 12 won by 30 percentage points. John believed that it became "one of the most profound electoral operations and should not ever be forgotten as one of the high watermarks of the wine industry's strategy of alliances with others."[21]

CONGRESSIONAL RECOGNITION

The next year John turned his political expertise toward the United States Congress. In 1990 Kiki de la Garza (D-TX), chairman of the House Agriculture Committee, suggested that the industry should push the House to establish a committee on wine. By forming a committee in the House of Representatives, the wine industry would have a vehicle to air their concerns directly with Congress and to bypass lobbying the bureaucratic ATF, which regulated the industry. After much interdepartmental haggling, de la Garza finally hit upon an approach whereby the wine industry would meet with the full committee on agriculture to address the following topic: "How could the Department of Agriculture further serve the interests of the wine and wine-grape industries?" In October 1991 representatives from different parts of the country testified before the committee. De Luca testified for the Wine Institute and in his testimony praised the body for their assistance on the International Market Promotion Program. As expected, the rest of his testimony centered on the topic of the health benefits of wine. John told members of the committee: "I think it would be a missed opportunity for the nation's health and for the nutrition of this country if in fact something as positive as a dozen studies talking about either the favorable attributes of wine or the cardiovascular benefits of moderate drinking—if we would have the USDA and this committee assist us, perhaps a modest program of some ten million dollars to further develop these studies."[22]

While his speech did not immediately convince many members of the committee to earmark money for research, it did help shift media coverage of the topic. Shortly after he testified, the CBS television program *60 Minutes* ran a segment called "The French Paradox." In the program, viewers learned that epidemiological data gathered by scientists showed that French people had a high rate of consumption of saturated fat and but less heart disease and a lower mortality rate than people in the United States. They labeled their conclusion the "French Paradox." After analyzing additional research, scientists attributed the positive paradox results to the French cultural habit of drinking red wine, which contains protective polyphenols and resveratrol from grapes. According to John:

> I certainly don't claim any influence on what *60 Minutes* did. As you know, their style if they are going to cover an industry they would go to the scientists, and they did with Curt Ellison of Boston College and Serge Reynaud from Lyon, France. We have not claimed that we had any impact on that program, but certainly the knowledge bank was there, and certainly there were researchers looking into this. It's a coincidence of history that we were testifying before the Congress for such a program, and then a month later the nation hears about it, which of course gave a lot of credibility to the request, because no one could ever claim 60 Minutes was bought or paid for by anybody.[23]

Needless to say, this was a major victory for the industry, as it proclaimed the advantages of wine as a food, a medicine, and a moderating beverage. Many in the industry wanted to immediately start advertising campaigns, which would include the use of positive labels, to let consumers know the health benefits of wine and to counter the damage done by mandated warning labels.

NEOPROHIBITIONISTS MAKE THE BATTLE PERSONAL

Despite the turn in scientific claims, neoprohibitionists continued their negative messages about alcohol. In 1991 California attorney general Dan Lundgren used Proposition 65 legislation to challenge the use of lead foils used for sealing wine bottles. Government lawyers claimed that more than six hundred wineries used lead seals, and that they presented a health hazard for pregnant women.[24] In a quick move to mitigate the issue, wineries settled the case by paying a $200,000 fine for failing to warn consumers and another $700,000 to update labels with a lead warning.[25] More important, in what seemed to be an overnight move, the industry abandoned lead foils.

It became obvious that much of the anti-alcohol energy seemed to emanate from rural and southern states. According to the publication *Wines & Vines,* US consumption of wine per capita had shrunk from 3.4 gallons per person in 1987 to 3.2 gallons in 1988. The industry publication claimed that the drop was a direct result of the neoprohibitionist movement in rural parts of the country. As evidence, they offered the observation that citizens in large cities like Washington, DC, drank a high of 6.2 gallons per capita compared to a low of 17.6 ounces per capita consumed by consumers in Mississippi.[26]

Despite the warnings and anti-alcohol admonitions, moderate drinkers began to think that many government agencies and policy makers were "systemically attempt[ing] to equate legal alcohol consumption with illegal drug use."[27] It did not help that Surgeon General C. Everett Koop declared that wine coolers were packaged to look like soda and that alcohol advertising to youths had to be restricted.[28] Many believed that the government had led a disinformation campaign against moderate consumption, and they worried that the lessons of Prohibition had been lost.[29] John Volpe of the National Wine Coalition warned: "You in the wine industry, you had better become activists very quickly because the other side is hammering you to death."[30] Many across the nation feared that wine would fall prey to the anti-alcohol jeremiad.[31]

Good news grew for those who believed that moderate wine

drinking was healthy as citizens digested a flood of new media information. In one news piece, a Florida reporter wrote, "Connoisseurs
are fighting for the right to drink in the face of anti-alcohol campaigns." The article continued with Julia Child saying, "'For centuries
and centuries, all, including Jesus Christ, drank wine. It's just part of
a civilized life.'"[32] Indiana reporter John Seiler vowed that authorities
would have to pry his fingers off the crystal goblet.[33] Famous Berkeley,
California, wine merchant Kermit Lynch pleaded for moderation and
believed that it was about "behavior control; it's always insidious the
way these tyrannical things come about."[34] Santa Cruz County pedia-
trician and winemaker Wells Shoemaker told a wine symposium that
"power rather than science and research" are behind the campaign
to make America sober.[35]

Many media articles began to attack neoprohibitionist actions
and policies. Readers in New Jersey read about how a Culver City,
California, school board removed *Little Red Riding Hood* from the
district's recommended reading list because it depicted Grandma not
only drinking wine but enjoying it.[36] Gene Ford, of Washington State
University, attacked anti-alcohol forces in his book *The Benefits of
Moderate Drinking.*[37] In 1990 K. Dun Gifford, a lawyer, restaurateur,
and chairman of the National American Institute of Wine and Food,
founded the nonprofit Oldways to promote healthy eating and drinking. Their work fitted perfectly into De Luca and the Wine Institute's
message of the food value of wine. The ties between food and wine
increased, and in 1991 the *Los Angeles Times* "Food Section" named
Robert Mondavi "Mr. Wine." *New York Times* wine writer Frank J.
Prial went as far as to suggest that a recent industry study showed
wine drinkers to be models of moderation.[38]

During this period, many anti-alcohol leaders referred to De Luca
as the "Booze Baron," and members of the industry began to feel the
personal pressure exerted by opponents. John remembered visiting
Christian Brothers winemaker Justin Meyer because he had become
disheartened over the name-calling. Meyer told him, "My kids have
come to me and my wife Bonnie and said that teachers had said that
those who were in the industry were baby killers and that people who
engaged in the wine industry were hurting society."[39] Dianne Nury,
first woman chairperson on the Wine Institute Board, said her children were getting the same message.

Wine-industry mothers Michaela Rodeno and Julie Johnson grew tired of the harassment and organized Women for WineSense to combat the negative name-calling. The two were taken aback in 1990 when their children came home from their Napa schools and asked them why they were "making drugs." It outraged the women, and they decided to fight back and counterbalance the accusations. Their goal was to find ways to discuss the benefits of moderate wine consumption and to create an image of wine as part of a healthy lifestyle. In a short time, other women in the wine industry, like Margaret Duckhorn, Rosemary Cakebread, Kit Wall, Lynne Carmichael, Dawnine Dyer, Zelma Long, Lili Thomas, Annette Shafer, Susan Sokol-Blosser, Gabrielle Saylor, Cathy Clifton, Margrit Mondavi, and many more, joined their cause.[40] Most important, the group caught the attention of the media and of state and federal legislators.

Further support came in 1993, when professors at the University of California, Davis, received a government grant to sponsor a seminar designed to bring U. S. scientists to UC, Davis, to discuss wine and health. At the get-together, John E. Kinsella, UC, Davis, professor of food science and technology, spoke about wine antioxidants and helped move the national conversation from the bad effects of ethyl alcohol to wine-specific studies that tied healthy antioxidants to the consumption of wine, fruits, and vegetables. There was now a direct correlation between what De Luca and the Wine Institute claimed and science that government officials could utilize for policies and legislation.

SEEKING CONGRESSIONAL ASSISTANCE

As the health issues related to wine slowly shifted toward the idea of a beneficial American wine culture, De Luca realized that he still had a long row to hoe in order to have the support of the majority of Congress. To move in this direction, he capitalized on work done by the Women for WineSense, who had organized and gotten legislative support for a California Wine Appreciation Week. Their initial work helped to convince Congress to sponsor a national event. John credited Women for WineSense for their key influence: "We needed a majority of both the House and the Senate, and the Wine Institute played an important role in this, particularly at the end when it looked doubtful that we could get the signatures. But the catalyst was Women for WineSense, and many other associations participated. They proclaimed the week of February 21–27, 1993, as Wine Appreciation Week."[41]

On Thursday and Friday before the Saturday start of Wine Appreciation Week, the political situation worsened for the wine industry. At a joint session of Congress, newly elected president Bill Clinton outlined his economic stimulus package, deficit-reduction proposals, and a health-reform package that included the words *sin tax*. Just as the Wine Week program commenced, many newspaper headlines capitalized upon the proposed taxes on wine, beer, and spirits to help fund health-care reform programs. De Luca went directly to the source of the controversy and remembered Clinton telling him that the idea of a sin tax was being promoted by others. Clinton then went on to say that excise taxes should only be on products, like tobacco, whose policymakers wanted to reduce consumption. But what struck De Luca and his staff was that the president said that he had not heard about any convincing evidence why there should be any increase in wine taxes. De Luca knew that the industry had some work to do if they wished to escape future anti-alcohol attacks.

De Luca helped orchestrate the week's events and purposefully planned an opening talk by Professor Dimitries Trichopoulos, a Mediterranean diet expert and member of the Harvard School of Public Health, who spoke about the role of wine in the diet. Dr. Lionel Tiger, a Rutgers anthropologist, spoke about historical uses of alcohol in America related to both in terms of puritanism and pleasure. He also spoke about how over thousands of years, the ancients had created a regime of food and wine and were cognizant of how wine interacted with the body, when used in moderation. Tiger also discussed the ethical issues of providing both positive and negative studies about alcohol and emphasized the public's right to hear both sides of the alcohol debate. That evening the Wine Institute sponsored a tasting and a screening of a movie for congresspeople, regulatory heads from numerous departments, the Food and Drug Administration, and the Bureau of Alcohol, Tobacco, and Firearms. De Luca's old friend Jack Valenti, now president of the Motion Picture Association, welcomed ten new congressmen and senators to Washington, DC, and then proceeded to introduce the evening's film.

Over the next few days, thirty guest speakers addressed the wine industry. People like Vic Fazio (chairman of the Democratic Congressional Campaign Committee), Leon Panetta (Clinton budget director), Bob Matsui (House Ways and Means Committee), Bill Thomas (California Republican congressman), Bob Torricelli (New Jersey Democratic congressman), and Senators Tom Daschle and Dianne Feinstein addressed participants. In between speakers and in the evenings, the Wine Institute delegation met with additional leaders, like House Majority Leader Richard Gephardt, Washington congressman Tom Foley, Lloyd Bentsen (secretary of the treasury), and Secretary of Agriculture Mike Espy. According to John, these were all meaningful encounters and not just social exchanges. They discussed policy and tried to be inclusive of both sides of the aisle as they approached both Democratic and Republican representatives and senators.

Additional programming included a panel discussion with Chris Matthews, an NBC television commentator and Washington Bureau chief for the *San Francisco Examiner,* along with Clinton campaign managers and political analysts Paul Begala and James Carville, who conversed about their views of the Washington, DC, press and media. On Wednesday evening, the institute sponsored a reception at the State Department with more than 350 guests, including 52 congresspeople and members of the administration. John reflected on the proceedings: "Congressmen and senators and members of regulatory agencies coming to a room where we were for the day, two days in a row, and on the half-hour, come and speak for about fifteen minutes on any topic: housing, health, international affairs, regulatory matters, food and drug, trade, what was going on with the new administration, plans for deficit reduction, economic stimulus. And then we would have time for opening up to questions from our audience."[42]

The audience for individual sessions included a delegation of forty Wine Institute members. More than thirty reporters covered the event's discussions on health, distribution theory, dissemination of information, legal issues, and ethical marketing. Most important, the event was the first inclusive, political get-together for the wine industry:

Also, in a very important way, the American Wine Appreciation Week events brought to Washington individuals other than the usual ones who came on the delegation. We had good representation from Women for WineSense, a good representation from the

American Vintners Association, the National Wine Coalition. There were individual delegations from New York, Oregon, Washington, and AWARE's president, Becky Murphy. We had representatives for the family winemakers. The CAWG group—the California Association of Wine and Grape Growers—and the Wine and Grape Growers of America, grape grower groups, also were in attendance. And therefore, in an unusual occurrence we had four or five times the number of wine industry representatives and leaders in Washington for a week in addition to the members of our own delegation. So this was probably in terms of the industry for the United States the single greatest attendance in Washington. And everybody shared in the information and shared in the news.[43]

In a few short years, De Luca had begun the process of shifting the health argument from the grips of neoprohibitionists toward the wine industry. He utilized third-party research, congressional support, the media, and appeals to a health-minded public to accomplish his goals. Public health and moderation now became talking points for the Wine Institute, and De Luca bombarded America with his message about the power of ancient food traditions and the need for a moderating wine culture.

{9}

Bionutrition, Pyramids, and Labels

This is the nation where we had Prohibition; this is not Europe, this
is not Italy or France, and I get back to the sense of promoting a diet
and a lifestyle, and this is exposing people to how the Mediterranean
peoples have lived and also the Asian peoples. Wine as food is a correct linkage, but this offers even greater opportunities, I believe. Wine
is part of the diet inherited by people.

—John De Luca, "President and CEO of the Wine Institute," oral history

As the barrage of attacks on alcohol shifted to defining wine as an
illicit drug, De Luca intensified his health and moderation arguments to
counter the neoprohibitionist doom-and-gloom rhetoric.[1] Anti-alcohol
proponents now condemned him as a purveyor of drugs and a threat to
the nation's health policy. De Luca responded by tightly clinging to his
personal beliefs in the tradition of wine as part of the Mediterranean
diet and the need for moderation as a means to resuscitate an American wine culture.[2] To show his concerns about alcohol abuse, De Luca
became a board member of policy groups like the Alcoholism Council
of California and the ecumenical North Conway Institute in Boston.

Making use of his past training, upbringing, science, and common
sense, De Luca took his narrative to policymakers and to the halls of
Congress, where he attempted to shift state and federal alcohol policies.[3] As part of this evolving political role, De Luca testified before the
Senate Subcommittee on Alcoholism and Drug Abuse in 1985. During
his testimony, he relayed the following to Chairperson Paula Hawkins
(R-FL) and the committee: "Your hearing provides the best vehicle to
rejuvenate our recent history where wine has been the principle beneficiary of the cultural movement in America emphasizing nutrition and
self improvement and the dynamism of women and the social emphasis
on beverages with less alcohol."[4] In a more scholarly approach, John
delivered his 1986 "Health and Safety Index" paper to the Washington, DC, Symposium on Wine, Health, and Society.[5]

Throughout the late 1980s, De Luca regularly reminded officials at the USDA about the nutritional value of wine and about how it could be included in their dietary guidelines. His persistence paid off in 1990, when Betty Peterkin, executive secretary of the Dietary Guidelines Advisory Committee, and Linda Meyers, executive secretary of the Health and Human Services Dietary Guidelines Advisory Committee, informed him, "The information you sent was carefully considered by committee members. We believe you will be pleased to see that their recommendations for the revised bulletin reflect your comments, at least in part."[6] Reporting on one such recommendation, committee member Malden Nesheim, PhD, informed Clayton Yeutter, secretary of the USDA, and Louis Sullivan, secretary of Health and Human Services, that the new dietary guidelines included the option of wine in moderation.[7] For De Luca, this was a major victory. Yet he knew that anti-alcohol forces would continue to fight this decision.

In his 1992 testimony before the Senate Committee on Commerce, Science, and Transportation, De Luca laid out how family, heritage, religion, and his Lower East Side upbringing influenced his approach to wine, health, and science. In a personal heartfelt plea, he stated: "I would like to establish a personal and cultural context for my remarks. My wife and I are both children of Italian immigrants, and terrible walking statistics in the minds of those who refer to wine as a gateway drug, or recreational drug, or drug of choice, or who deliberately blur the distinction between a glass of Chardonnay and cocaine, a glass of Zinfandel and crack."[8] He then went on to inform committee members about how he grew up with a wine culture, took wine at Mass, made wine with his father and brother, and spent a lifetime eating a Mediterranean diet that included wine in moderation.

John was not alone in his defense of wine, and many defended his message of wine moderation and health. Mark Fisher, reporter for the *Dayton (OH) Daily News,* agreed when he wrote, "It appears common sense is among this war's earliest victims." But he also added the caveat that the wine industry initially overreacted to the anti-alcohol attacks. As a wine drinker, Fisher supported the industry's biblical, cultural, political, and Jeffersonian calls for wine temperance, and he sternly reminded readers that Prohibition had failed. In his commonsense approach, he pleaded, "Those whose voices need to be heard right now are the consumers. Only we can protect ourselves from unnecessary infringement on our rights, while recognizing some political and social realities."[9] De Luca agreed with Fisher's analysis, and over the next decade he navigated between hostile forces, on both sides, to achieve a middle ground for an American wine culture.

THE NEW SCIENCE OF BIONUTRITION

Since 1916 the United States Department of Agriculture had offered a list of nutritional suggestions for the American people, and by the 1930s the guidelines included a list of foods categorized by food groups with measurements for amounts to be eaten. The USDA updated this approach during the 1940s by organizing a complex system built around goals for nutrient adequacy derived from seven food groups. In an attempt to make the recommendations more accessible, they simplified this approach in the decades of the 1950s–1970s by reducing the number to four food groups and by recommending a moderate intake of fats, sweets, and alcohol. As nutrition science advanced, the USDA updated its guidelines by incorporating a 1980s "Food Wheel" and a total dietary approach that included calorie-intake suggestions from five food groups. But it would be new research in the late 1980s by scientists like Linda Bisson, professor of viticulture and enology at UC, Davis, that really increased interest in the new science of bionutrition and which translated the knowledge of clinical science into basic suggestions for how to improve nutrition.

As bionutrition and research into the Mediterranean diet progressed, De Luca saw an opportunity to reframe the role of wine in American society. The 1990 USDA dietary guidelines included a recommendation of moderate wine consumption for those eating foods based on a Mediterranean diet. For many wine enthusiasts like De Luca, this shift toward accepting wine as a part of a balanced nutritional program brought hope to the cause of rebuilding an American wine culture. Increased acceptance surfaced in 1992, when scientists suggested a new visual tool, a pyramid of suggested foods. The new chart recommended a total dietary approach built upon research showing the need for variety, moderation, and proper proportions. USDA scientists had borrowed the idea from the 1972 triangular pyramid that had been issued by the Swedish National Board of Health and Welfare.

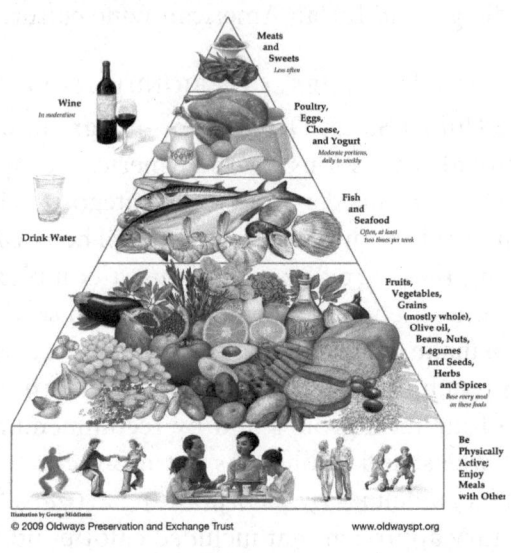

The Mediterranean Diet Pyramid, 2009. Courtesy of
the Oldways Preservation and Exchange Trust.

The plan's serving suggestions included six to eleven servings of grain
products; two to four servings of fruits and vegetables; two to three
servings of meat, poultry, fish, and beans; and at the top of the pyramid, moderate amounts of red meat, sugars, and fats.[10]

As the nutrition movement grew, many universities, colleges, and
businesses increased their research on nutrition and diet. In 1993, an
International Conference on the Diets of the Mediterranean was held
in Cambridge, Massachusetts. Organized by Oldways, a nonprofit
food and nutrition organization, and the Harvard School of Public
Health, the conference brought together scientists, chefs, food writers,
and experts to present the latest nutrition science around the Mediterranean diet. One important result was the introduction of the Mediterranean Diet Pyramid, an alternative to the USDA Pyramid, and a
graphic representation of the foods that make up the diet—primarily

whole grains, fruits, vegetables, beans, nuts, legumes, and olive oil. For the first time, moderate wine consumption and physical activity were included along with the following elaboration: "Following Mediterranean tradition, wine can be enjoyed in moderation, primarily with meals; one or two glasses a day for women and men respectively. It should be considered optional and avoided whenever consumption would put the individual and others at risk."[11]

An encouraged De Luca considered this shift in policy to be a breakthrough moment and later commented:

> I see that as a much more dynamic concept, wine as part of the diet of people, than the familiar one of wine as food, concept, wine as part of the diet of people, than the linear one of wine as food. Because diet connotes a whole body of thinking of a whole category of people over time rather than an individual by individual choice. So, I think we can counter and address these two major issues simultaneously to attack those who want to make us a gateway drug, a sin to be taxed, and to counter their perception that we're a quasi luxury product by in fact indicating that wine has been for thousands of years, for hundreds of millions of people, part of their staple of life and of their diet—along with their pasta, along with their olive oil, along with their wheat bread was a glass of wine—and get it back to that perception.[12]

It proved to be an important convergence of causes, as the Harvard School of Public Health and the United States Department of Agriculture agreed to the wine/food message. But in order to take advantage of the situation, De Luca needed congressional appropriations to fund additional research. To do so, he worked with the USDA to encourage various departments to fund research projects investigating nutrition and moderate wine consumption. To this end, De Luca appeared before the House of Representatives Committee on Agriculture. In his testimony he told congresspeople that the "government has only hurt us when it's been uninformed. We do not fear informed government, but this century has been marked by the enormous ravages that have been really the historic impediments to our growth." De Luca ended his testimony by challenging Congress to investigate and sponsor additional research into alcohol abuse and moderation.

"It is my sense that it is very appropriate that the Health and Human Services issue grants that look into abuse, but I think it would be a missed opportunity for the Nation's health and for the nutrition of this country if in fact something as positive as a dozen studies talking about either the attributes of wine—resveratrol, quercetin—and the cardiovascular benefits of moderate drinking—if we would have the USDA and this committee assist us in a modest program of some $10 million to further develop these studies."[13]

Over the next few years, Congress released funds for research, and the Human Nutrition Research Center in the Department of Agriculture funded six different projects on energy metabolism and caloric intake. Further good news came in 1994, when the USDA created the Center for Nutrition Policy and Promotion to improve the health and well-being of Americans by establishing national dietary guidelines based on the best science available. Attempts to get third-party research money had succeeded, and in a 1996 "Presidents Report" to Wine Institute membership, De Luca announced that NIAAA had awarded fifteen grants totaling more than $2,740,577.[14]

De Luca also began contacting the Department of Health and Human Services, calling on them to counter the persistent anti-wine bias promoted by anti-alcohol forces. He approached Dr. Philip Lee, assistant secretary of Health and Human Services, in the hopes of working with the National Institutes of Health, the Food and Drug Administration, the surgeon general, and other agencies. De Luca visited Lee and laid out his proposals for new legislation and asked for criticism and input so that no one would be blindsided.

For the good part of a year, De Luca and Lee, along with Linda Bisson at UC, Davis, strategized how to encourage the federal government to promote bionutrition as a policy direction for nutrition, diet, and lifestyle issues. They made sure to include roles for the National Institutes of Alcohol Abuse and Alcoholism and the National Heart Association. John also visited Dr. Enoch Gordis (director of the NIAAA), staff members Christy Carpenter and Elizabeth Holmgren, and ten NIAAA division heads. As a result of the meetings, De Luca felt that he had convinced the NIAAA that the Wine Institute would support only good, refereed science. Most important, De Luca kept in constant contact with Lee while writing the report. He remembered:

Week after week he would send messages over the fax with his additions, I would send over my additions, I would talk on the phone, I would go to Washington, we would have conferences. I would meet with Dr. Lee here in San Francisco when he came to San Francisco—he resides here; he used to be the former chancellor of UC, San Francisco. His office would keep me informed of his travel plans, and if I knew I was going to be in Chicago and he was going to be in Chicago, I would try to set up meetings. So by phone, by fax, in person, by working with his staff, we slowly over a ten-month period came to a polished version of what we as the Wine Institute would then present to legislators.[15]

When it was time to present their findings, the group sought the assistance of Representative Nancy Pelosi and Senator Dianne Feinstein, both of California. De Luca, Lee, and Bisson knew that their proposals were a hard sell because they asked the government to use federal taxpayer money to do research on the benefits of drinking alcohol. But with the support of top health officials, the political savvy of Pelosi and Feinstein, and support from Representatives Steny Hoyer (D-MD), Dick Durbin (D-IL), Anita Lowe (D-NY), and Rosa De Lauro (D-CT), they secured funding for numerous appropriation bills. De Luca said, "It was one of the great ironies of my career here that twenty years later, the public's right to know, which was kind of the shibboleth used by the Center for Science in the Public Interest in the early seventies about ingredient labeling, would somehow come back and be repositioned where we would be in a position to say, What are you afraid of?"[16]

Initial good news came when Dr. Gordis thanked De Luca for his input and confirmed that "the National Institute on Alcohol Abuse and Alcoholism is interested in supporting research."[17] De Luca also got a confirmation letter from Secretary Lee. In the letter Lee wrote, "I am pleased that the NIH will be increasing its support for research into the health effects of moderate drinking if Congress supports the Administration's request for NIH in the fiscal year 1996."[18]

Congress approved the additional funding for research grants, and scientists were invited to apply. But it became apparent that top scientists hesitated to become involved in what they perceived to be politicized alcohol research. They did not want to appear to be doing promotional work for the wine industry. In response, De Luca met with members of the research community to alleviate any fears they might have about being used as a marketing tool for the Wine Institute. Fortunately, John was able to convince many of them.

> Any idea that what they were doing could be used to "push alcohol," in their minds, would be anathema. But I think they recognize—because there were bruises and scar tissue on me—that this was not in fact the case. And they told me that. They said that if it had been presented any other way, if there had been one scintilla of doubt that what we were doing was in fact a front for some of the wineries' marketing programs, they would have walked away from us. It was that tenuous a relationship.[19]

At this time it also became apparent to De Luca that media coverage would be necessary to disseminate research results to the public. "Because we are so health conscious, we will eat a product or not drink a product depending on what was in *Redbook, Vanity Fair, USA Today,* and CNN. And so the interaction between the scientific community and the media seemed to me to be the most legitimate path to take, and my thinking here is that we cannot be the ones to proclaim it; we need third-party credibility."[20] But for De Luca, this type of promotion presented itself as an ethical balance among third-party research, promotion of a positive wine message, and responsiveness to those suffering from alcoholism. De Luca later reflected:

> What it really meant was a balancing act between these two requirements, very legitimate. If anything, it was a dilemma. It wasn't anything malicious. It's do you wrestle with this major issue in modern times. I think we have bridged it in a very significant way here at the Wine Institute. The Wine Institute's role has been in some cases instrumental, in some cases has been augmenting, to disseminate the information without running afoul of the government agencies or the legal profession. And out of this period of tension, I think, has emerged a very, very good approach, is that the Wine Institute and its members as individuals should disseminate information that is the consensus of the scientific community in terms of the marketing of ideas, rather than trying to market a product.[21]

To achieve his media goal, John turned to his White House Fellows classmate and friend, Tom Johnson. After working in the Lyndon Johnson administration, Tom served for ten years as the publisher of the *Los Angeles Times* and in the 1990s became the president of CNN. John asked his longtime friend for advice:

> As you know, wherever you travel, people turn on TV and there's a CNN story. So I recognized very early that we should establish our credibility with them, and I wanted to know who some of their people were, and through Tom Johnson's office met with the nutrition and features people, the people who are covering foreign affairs, and particularly a unit headed by Carolyn O'Neil which does a program, "On the Menu" that covered the Mediterranean diet better than anybody else had, with very wonderful footage and a very good interpretation. They have done a lot of pieces not only on the Mediterranean Diet, but on the cardiovascular benefits of wine, and on just wine stories in general.[22]

Further success for the food and wine cause came in 1995, when the US Dietary Guidelines acknowledged the health effects of moderate wine and alcohol consumption. Under the new guidelines, the revised Mediterranean Diet Pyramid presented a balanced position on wine that described the possibilities of reduced risk for heart disease, as well as the risks of abuse for some individuals. As it had in past food guidelines, the official nutrition policy of the federal government acknowledged that alcoholic beverages have enhanced the enjoyment of food in many cultures and that there are cardiovascular benefits provided by drinking in moderation. De Luca had successfully countered previous language in government policies that stated that wine had no health benefits and which linked wine to illicit drugs.

Moderate wine consumption now had the support of government policies, and De Luca believed that this was a "major turning point, because it shows that when you have wine with meals, you're not off by yourself, you're not drinking to just drink, that it's part of a social setting or family setting."[23] The new direction however, did not please everyone, and in a *San Francisco Examiner* article, Hilary Abramson, former staff member for the Marin Institute for the Prevention for Alcohol and Other Drug Problems, warned De Luca to be careful about the policy win by reminding him that the Wine Institute was a promotional organization and that this move could result in wine being transferred from Alcohol, Tobacco, and Firearms to the Department of Health and Human Services, and insinuating that wine would be regulated like any other food or drug.[24] The veiled threat did not dampen De Luca's enthusiasm, and a few days later he responded to Abramson by saying that "the impact of the new wording was so positive that the wine industry might help distribute the new guidelines."[25]

By this time, De Luca and the Wine Institute had become the industry's premier wine organization and a political powerhouse with a staff of more than 100 people, 42 offices, and 430 paying members (out of a total of 740 wineries), and they represented 90 percent of America's total wine production. To ensure future political support, the Institute's bipartisan political action committee donated to wine-friendly candidates on both sides of the aisle. In the 1996 election cycle, the institute distributed about $100,000 to campaigns, including $8,500 to Senate Democratic leader Tom Daschle (D-SD), $7,000 to Representative Frank Riggs (R-Windsor), and $1,500 to Representative Lynn Woolsey (D-Petaluma). Individual wineries also contributed to selected groups and candidates.[26] When the Wine Institute spoke, government agencies, organizations, and anti-alcohol proponents listened.[27]

The next year the prospects for continued wine acceptance grew when the Wine Institute advocated for and won more than $10 million for fifteen bionutrition research projects sponsored by the NIAAA. Many of the projects targeted research on moderate alcohol consumption and health, and by 1997 seventeen government-financed and -moderated grants totaling $2.7 million per year for five years were in force. John gave much of the credit to Wine Institute Washington, DC, lobbyist and office director Bobby Koch, as well as to Representative Nancy Pelosi and Senator Dianne Feinstein, for their political savvy in making it happen.

DIRECTIONAL LABELS—NEOPROHIBITIONISTS BATTLE HEALTH CLAIMS

Needless to say, the anti-alcohol forces had suffered a setback in their plans to restrict and demonize all alcoholic beverages. But they were far from giving up their cause and began crossing swords with those who touted the health benefits of wine.[28] More than ever, they returned to their core value of playing on emotionalism when discussing the ravages of alcohol abuse. They used terms like *gateway drug* and *sin taxes* to put wine on the defensive as they attempted to make wine-supportive politicians, researchers, media officials, and decision-makers appear to be evildoers. These continuing attacks scared many wine pioneers, who remembered the "dry times."[29] John believed that anti-alcohol followers "have done great harm to the American understanding of the food chain; where we have the highest and safest standards in the world, and yet more and more American consumers are confused as to the healthfulness of their own products."[30] The reality was that all alcohol had been given a negative image by many federal agencies that had systemically equated legal alcohol consumption with illegal drug use.[31]

John retrenched and settled in for the long haul, in what he called the philosophy that "long journeys begin with an initial step."[32] He knew that it would take ten to fifteen years to get adequate research funds and final results from completed projects. Knowing that he needed to maintain his moderation message, he philosophized:

The only way to counter it is not to rail against it but to provide facts to answer the questions with medical consensus. That is the approach that I believe we've taken. In American democracy everybody can participate in the public policy debate. Whether you have the credentials or not, whether the press finally cuts through it all and says, "Well, wait a minute. You're not in peer review journals. Where did you get your information?" There is a certain element of skepticism. I have no personal problems or animosity toward any of these groups. I think that's the price you pay for democracy in America. What I'm disappointed in is the inability of certain public officials and certain reporters to understand that even though they say their cause is correct, in terms of fighting abuse, they really haven't focused the case on abuse. They really have shifted the debate toward use, consumption. And that's where I think they have unmasked themselves. If only the public again, the media, here, has a responsibility. We're sophisticated and knowledgeable enough—and I attribute this to cultural ignorance in the United States, not to anything malicious—to understand that the industry itself makes common cause with those who attack the abuse of the product. And what is nefarious, what is intellectually dishonest, is really not permissible in terms of research is to make this leap from the abuse of product and all that follows, to say that that is also the weight of responsibility on moderate behavior.[33]

In 1995 anti-alcohol forces fired their next salvo by challenging the wine industry's proposal for a wine-bottle front or back directional label that would use scientific information to help contextualize and balance negative warning labels. Since the *60 Minutes* episode, wineries had been pressuring the Wine Institute to convince the Treasury Department to allow them to mention wine's health benefits on labels. In response, the Wine Institute had been working on submitting a new type of label that described both the health effects of drinking in moderation and the effects of alcohol abuse. De Luca also felt that this type of language would quell congressional concerns about fair and balanced consumer information and at the same time highlight the need for further research about the health effects of drinking in moderation. De Luca assured everyone that the proposed label, was voluntary, was meant only as a supplement to the warning label and would give consumers a more scientifically balanced view of the product.[34] Dr. Philip Lee agreed and went as far as to say, "In my personal view, wine with meals in moderation is beneficial. There was a significant bias in the past against drinking."[35]

As the industry presented the label to the Treasury Department, anti-alcohol forces pushed back. The Center for Science in the Public Interest claimed that the new label would encroach upon the warning label mandated by congressional law. The CSPI maintained that the intent of negative alcohol and tobacco warning labels was to improve public health by keeping many consumers from purchasing the product. George Hacker, CSPI director of alcohol policies, believed that the proposed directional labels would mislead consumers and promote abusive drinking.[36] Therefore, the CSPI confirmed that they could not agree to any label with a moderation message because they believed that you cannot drink an inherently dangerous product in moderation. The group then threatened legal actions like those experienced by the tobacco industry to ensure that the warning label remained and that there be no hint of a moderation message. Many agreed with the

CSPI stance, and in a *San Francisco Examiner* editorial, the newspaper's board opined that the health claim could not be justified, considering that one hundred thousand Americans died yearly from alcohol abuse. They went on to satirically nominate the Wine Institute "for the Emperor Norton Prize by seriously urging the government to imply on the label that wine is somehow a medicine."[37]

Given the intensity of pro and con arguments, the Treasury Department cautiously ruled that wine labels were too small to include all sides of the issue and that incomplete messages could possibly interfere with the intended message of the warning label. Not everyone was willing to throw in the towel, however, and the Competitive Enterprise Institute, a free-market think tank, filed a lawsuit against the ATF, arguing that consumers had a right to know the health benefits.[38] The CEI lost the case, but the issue was far from dead.

The momentum against advisory labels shifted in 1995 when the Department of Agriculture's *Dietary Guidelines for Americans* stated that moderate drinking could be associated with a lower risk for coronary heart disease for some individuals. It was now impossible for the ATF to ignore the issue of directional labels that were promoderation and made health claims. To cover their bases, the Center for Substance Abuse Prevention (CSAP) in the Department of Health and Human Services found a middle ground to the controversy by calling for field surveys to establish what consumers' reactions would be toward the new advisory labels. In response, De Luca signed a release so that the Wine Institute's confidential and proprietary label could be publicly discussed. Nationwide, scores of newspaper editorials supported giving more information to consumers, and most agreed that the absence of the new directional label dumbed down consumer awareness of current scientific information.

On the other side, anti-alcohol groups quickly responded to the call for field surveys and attempted to use it to their advantage. CSAP believed that the surveys would vindicate their point, and in a behind-the-scenes move they—without the wine industry and Treasury's knowing about it—commissioned independent researchers to do face-to-face interviews with consumers. The researchers asked 404 consumers in four cities across the country what their response was to the new advisory label. Then they brought together two focus groups in the Washington, DC, area and collected their reactions to the proposed labels. The CSAP's tactic backfired, as most consumers did not see the label as a marketing ploy and knew the difference between the old and new labels.

On the basis of the survey, the Treasury Department surmised that the new advisory label did not mislead consumers and in 1999 approved two well-crafted labels that read:

> The proud people who made this wine encourage you to consult your family doctor about the health effects of wine consumption.

> To learn the health effects of wine consumption, send for the Federal Government's Dietary Guidelines, Center for Nutrition Policy and Promotion, USDA, 1120 29th Street, NW, Washington D.C. 20036 or visit its website.[39]

By the end of 1999, the ATF had approved ninety-nine wine labels with health-related advisory statements. For De Luca, "The label approval represents a defining new chapter in the evolution of federal policy toward wine in America."[40] In a special faxed memo, John told the Wine Institute membership, "The label approval is a stunning victory for our public policy goals."[41]

The success of the new label angered and motivated Senator Strom Thurmond to retaliate, and in a short time he lived up to his *Wine Spectator* title of "Wine's Public Enemy #1."[42] As part of his new anti-alcohol campaign, Thurmond introduced three bills to attack wine. He wanted to triple the federal excise tax on wine and use the estimated $7.9 billion to help fund the National Institute of Alcohol Abuse and Alcoholism. The other two bills would ban all health claims on labels and move alcohol from the ATF to the Department of Health and Human Services.[43] Thurmond also threatened to hold up future Treasury Department appointments until his bills passed.

A series of contentious discussions with the FDA, CSPI, Thurmond, MADD, and the Wine Institute followed. Many warned that health claims could result in wine being labeled as a drug, necessitating that it be moved from ATF to FDA jurisdiction. The CSPI, MADD, and Thurmond argued that the new advisory health labels were nothing more than governmental approval for drinking.[44] In a letter to Roger C. Viadero, USDA inspector general, Thurmond harshly criticized the Wine Institute and De Luca for trying to undermine American dietary standards through the power of the alcohol warning labels. He claimed that the Wine Institute had "used" and "manipulated" many members of the Guideline Committee, and he condemned Phil Lee for omitting drug references about wine.[45]

Thurmond's pressure on Secretary of the Treasury Robert Rubin worked, and as a compromise, Rubin agreed to put out a notice of proposed rulemaking that would elicit comments from the public. In return, Thurmond helped push his three appointees through the confirmation process. De Luca feared that this new direction would create "litigation, class-action lawsuits, either, that would pursue wine, beer, and spirits the way they've pursued tobacco. Those are not marketing issues, those are public policy issues."[46] Following his past strategies, De Luca took to the airwaves and appeared on the Bay Area's channel 4 KRON *Today Show* to debate Sarah Kayson, the public policy director for the National Council on Alcoholism and Drug Dependence. Kayson doubled down on the idea that heavy drinkers would equate health benefits with heavy drinking and praised Strom Thurmond for his new bills aimed only at wine because the industry had made health claims.[47]

John reflected on the heated advisory-label battle:

Our opponents, who have tried to keep this from happening [directional labels] have tried to derail it and are now trying to diminish its success by saying, "Well, that's no big deal. This directional label doesn't mean anything. It's so watered down it's not even important." That's a time-honored tactic: if you cannot stop something, you try to belittle it. The proof of that historically, of how important this battle has been, is that it was in May of 1996 that we offered it, and here we are March of 1998—more than twenty months away—and it's been fought mano-a-mano like foot soldiers. That should give you an idea of what our opponents thought of it.[48]

De Luca acknowledged that the battle for an American wine culture was far from over.

Direct Shipping

By 1985, we found that the increase in the number of wineries and consolidation of the number of wholesalers led to a number of wineries not being able to find wholesalers to carry their product. They weren't attractive enough in their portfolio. So they came to us. I remember a number of meetings. And said, "You know, we're being excluded from our own markets in our country. We have people who come to our wineries who want to buy our wine, and they can't find it in their own home state. We can't even ship it to them."

—JOHN DE LUCA, "President and CEO of the Wine Institute," oral history

NEOPROHIBITIONISTS CLING TO ARCHAIC LAWS

The repeal of the Prohibition Amendment left the United States with fifty sets of state laws that restricted the sale and distribution of alcoholic beverages. This was no coincidence but a measured move by neoprohibitionists to keep control of America's drinking habits. When Franklin D. Roosevelt ran for president in 1932, he proposed a platform that included a plank endorsing an amendment to repeal Prohibition. To achieve this goal, he needed to have two-thirds of the Congress and three-fourths of the states ratify the amendment. The new president understood the ongoing political battle between wet and dry proponents and the disconnect between urban and rural citizens as they battled for control of the federal government. In order to get a repeal amendment, Roosevelt negotiated a compromise among states, anti-alcohol forces, producers, and alcohol-friendly citizens. In order to get the ball rolling, Roosevelt proposed empowering states with the right to control alcohol production, sales, and distribution within their own boundaries. FDR then pushed the deal over the finish line by promising state governments control of alcohol excise taxes to help offset state budgets ravaged by the Great Depression. His proposed Repeal Amendment (the Twenty-First) passed Congress and the states in 1933, and the United States subsequently developed an alcohol-distribution system divided between controlled states and free states.

THE THREE-TIER SYSTEM AND RECIPROCAL TRADE

A by-product of the Repeal Amendment was the evolution of the three-tier distribution structure that replaced the previous system wherein producers could have control over shipment as well as retail sales by owning both wholesale and retail outlets. After Prohibition many anti-alcohol politicians attempted to control all licenses for the wholesale and retail distribution of all alcoholic beverages. To do so, they established laws that directed producers to sell only to state-licensed wholesalers, who in turn sold to retailers, who in turn sold to consumers. From this evolved the present-day, three-tier system of producer, wholesaler, and retailer. In the beginning eighteen states chose to be control states, serving as the wholesaler, as the retailer, or as both, rather than leaving these roles to the private sector. Through the years, these laws became even more entrenched, as some states created franchise laws that gave private wholesalers a monopoly. As a result, many state governments became involved in what would otherwise be the functions of private businesses, and wine was not considered to be a free-trade item. In reality, the wine industry had been stripped of all protections under the United States Constitution's Interstate Commerce Clause. For De Luca, the opening up of markets in the different states and dealing with their regulatory agencies and legislatures became prime concerns.

As the three-tier system expanded over the second half of the twentieth century, most large production wineries utilized wholesalers to distribute their wine throughout the national market. But as the wine industry consolidated, so did the number of wholesalers, and by 1985 many limited-production small wineries found it difficult to find national wholesalers willing to carry their products.

After receiving complaints about restrictive domestic markets, the Wine Institute leadership prioritized direct sales as a top actionable issue. Small wineries faced a dwindling number of wholesalers and viewed direct sales as a new niche market and as an opportunity to be competitive with large wineries that had access to the three-tier apparatus. As expected, private and state wholesalers hollered foul because they felt direct sales circumvented the established three-tier system. Complicating the situation was the fact that most control states rarely enforced misdemeanor alcohol laws, and distributors feared that they had few legal options if small wineries broke state laws.

The real question was what solution would solve these problems for all concerned parties. The wine industry had no overarching policy, and this led some wineries to call for a constitutional amendment to dismantle the three-tier system. Many other wineries just continued shipping wine under the radar, and large wineries continued as usual with the three-tier system.

Initially, De Luca felt that the best way to approach these conflicts was to enter into reciprocal trade agreements with as many states as possible. The intention was to allow member states to trade wine freely across their state lines. De Luca understood that repeal had given states the sovereign power in alcohol matters and approached the issue as if it were an international trade negotiation. He told his membership, "Let's engage in trade, let's have trade agreements between California and Oregon, California and Nevada, and California and Colorado. Let's have reciprocal trade agreements."[1] In 1985 De Luca pushed to convince the California Legislature to pass legislation for reciprocal trade agreements with individual states. Washington and Oregon immediately signed the agreement, and in a short time the list grew to fifteen states. The intention was not to bypass the three-tier system but to provide a safety valve for small wineries so that they could work through the restrictive process.

As winery tourism grew in popularity, most large and small wineries offered tasting-room direct sales and wine clubs that shipped their wines to customers in state and out of state. Even high-end men's haberdashery Brooks Brothers started a wine club with direct sales.[2] By the end of the nineties, a multitude of wine clubs relied upon direct shipping of telephone, fax, and internet orders. The new trend alarmed wholesalers, who saw this as a direct threat to their livelihood.

Tensions increased as many wineries flaunted misdemeanor state laws by shipping wine directly to consumers in states without reciprocal agreements. The loopholes in the new direct-shipping schemes increased wine sales, but they also caught the eye of anti-alcohol groups who believed that the trend would result in a rise in alcohol abuse. In order to push back against direct sales, anti-alcohol supporters and wholesalers, in yet another "Bootleggers and Baptists" scenario, joined together and lobbied state legislators for laws that would permit the use of federal courts to enforce state alcohol laws. Since most of the state alcohol laws were misdemeanor crimes, they carried minimal punishments that many consumers and sellers ignored. To enhance enforcement of the laws, many states and distributors now pushed to make it a felony to ship to nonreciprocal states.[3] For most members of the wine industry, a felony referred to crimes like murder, arson, robbery, and assault, and they feared that a felony conviction on their record could cause them to lose their business license. Unfortunately, the whole issue escalated and resulted in government hearings and press conferences aimed at demonizing the wine industry.

SAVING THE CHILDREN FROM THE EVILS OF ALCOHOL

Both sides strategized and proposed regulations to oversee the direct shipment of wine. Wholesalers attempted to convince Americans that kids would be able to purchase alcohol illegally and that states would lose tax revenues if direct shipping were made legal. It did not take long for the Wine and Spirits Wholesalers of America to realize that their ally in the direct-shipment battles were religious neoprohibition-ists and groups like Mothers Against Drunk Drivers, Students Against Drunk Drivers, and various alcohol-abuse organizations. Driven by the common goal of wanting to keep the three-tier system as a means of regulating alcohol distribution, the new anti-alcohol coalition set to work. In the South a number of wholesalers and state organizations pressured their legislators to enforce three-tier laws, and in a short time Kentucky, Georgia, Florida, and Tennessee passed legislation that made it a felony to ship beer or wine directly to consumers. De Luca recounted meeting with the governor of Georgia and members of the state legislature and asking: "There's got to be a more intelligent way to handle this. It's not inside baseball. It's not just between the tiers. It's trying to respond to consumers. Consumers in your own home state can't find these wines, so they are asking us to send wine directly to them, and the wholesalers don't carry the wine. So effectively they're being barred from doing commerce in your 'state.'"[4] Throughout the discussions, state officials emphasized their Twenty-First Amendment power to control alcohol sales and distribution.

Chicago-based Beer Across America complicated the problem by beginning a nationwide program for direct retail sales of beer. This angered many state attorneys general because, just as with wine sales, these beer sales impacted individual state three-tier monopo-lies. Most states found it difficult and financially costly to enforce their misdemeanor laws, but this did not stop many district attorneys from organizing sting operations and from encouraging media outlets to use investigative journalists to bring attention to violators. In a short time, seven states passed felony direct-shipment laws and began the process of prosecuting consumers for receiving direct shipments purchased at a winery or through telephone, fax, and internet sales. Attorneys general, state legislatures, anti-alcohol groups, and wholesalers all protested the use of technology for interstate commerce to circumvent the regulated, person-to-person three-tier system.

So just at a time when consumer demand rose, wholesalers in the three-tier structure fought and threatened alcohol producers. Over the decades, the Wine Institute had done a large amount of work to ease the political, economic, and legal problems related to the internal trade problems created by repeal. At the Wine Institute, De Luca found himself caught in the middle: "We had a portion of our industry that was not reaping the benefits of the new consumer interest because of the system in place; but, I had members in place who had no beef, had no animosity toward anybody."[5] His political training had taught him that America was not only a political union of states but also a common market that protected interstate commerce through the Constitution's Interstate Commerce Clause.[6] The political battle for an American wine culture had shifted, and now many states ignored the Constitution's Commerce Clause for the Repeal Amendment's promise in favor of controlling alcohol issues and revenues within state borders.

Reciprocity had not achieved the solution needed for wineries, and De Luca now turned to negotiating with eighteen control states to further open the market. For many years, the Wine Institute hired state-level contract lobbyists in an attempt to roll back provisions in state laws by promising states their fair share of excise tax money and license fees. In New Hampshire, Nevada, and the Dakotas, De Luca promoted the idea of special personal permits that would allow people to buy wine directly. The plan included maximum purchase amounts, and all direct deliveries required adult signatures. De Luca promoted the transparency of the negotiations through national media outlets like *USA Today* and *The Wall Street Journal*.

Despite the institute's aggressive work, more than a dozen states had established felony laws designed to punish those engaging in direct sales. At the Wine Institute, De Luca and Steve Gross, the institute's director of state relations, began talks with these states in attempts to negotiate regulations favorable to both sides. The two visited Kentucky three times to meet with the governor, pleading the case for Kentucky visitors to California being allowed to buy wine at a winery and ship it back home. In Louisiana the two got a compromise that provided for sales licenses and collection of state revenues. The silver lining to the shipment chaos was the fact that it helped unify the industry by giving wineries a common enemy. Groups like the Family Winemakers of California, the American Vintners Association, the Coalition for Free Trade in Licensed Beverages, and the Wine Institute joined together to form the Direct Shipping Coalition.

After releasing a scathing condemnation of internet wine sales, Senator Orrin Hatch (R-UT), chairman of the Senate Judiciary Committee, held a hearing on the issue that included testimony from De Luca.[7] During the hearing, Hatch featured a Washington State thirteen-year-old girl who had participated in a sting operation and used her parents' credit card to purchase and have beer delivered to her home without any request for an adult signature. Anti-alcohol forces latched on to the case and cried foul over what they saw as a marketing practice of selling to minors. Wholesalers joined with anti-alcohol forces and nicknamed the ability of consumers to directly purchase alcohol through telephone, fax, and online as "cyber booze." Seeing this as an opportunity to stifle direct sales, they argued that they were there to protect American children. De Luca's response to their new rhetoric was as follows:

> Look, that isn't the way to go about it, attacking each other. We're talking about which system is doing this. That's not the issue. We have very strong social responsibility issues here. We don't practice marketing to minors and we don't use athletes. We have denied ourselves some of the marketing practices that would be available in a free society, because we don't want to have those heroes of the young, whether they're stars or athletes, and so we can make the case that far from us trying to contribute to the delinquency of minors, we really have been a force in the other direction: how we advertise, how we educate, our educational programs.[8]

Despite the growing animosity between the wine industry and

wholesalers, De Luca remained levelheaded and soldiered on to resolve the underlying problems. He later recalled how he felt at the time:

> Black marketeers: we haven't heard that in sixty-five years; that we contribute to youthful violence, which is nonsensical; cyber-booze—all these epithets have been hurled at us. But we feel somebody has got to be a steward of this reservoir of social responsibility that all have created, including wholesalers, and we don't want to dip our hand in mud and start throwing it at them. So even though some of our most vocal adherents want us to get in there and have knives in our mouth and climb the riggings and go after them, we've tried to keep the issue focused because we need to continue as partners. If there are going to be eventual resolutions of these issues it has to be negotiated. And we all should address the underage issue, but electronic commerce, you shouldn't deny a budding industry, and particularly to those who by virtue of the law don't have access to the consumer. Remember, if you're a three-tiered state, and you must go through the wholesaler but the wholesaler doesn't carry, for whatever reason, how are you to engage in commerce? And if a consumer can't find your product, but hears about it, reads about it in many of these wine journals and writes to you or calls to you or finds a website on your page and asks you to send it directly to him or her, should that be a felony that puts you out of business?[9]

With lobbyist support from the Wine and Spirits Wholesalers of America, Senator Hatch and Senator Mike DeWine (R-OH) joined together in a crusade to keep teenagers from having access to alcohol. They also proposed federal legislation that would make it illegal for wineries, retailers, and wholesalers to offer internet sales and would require enforcement through the federal courts.[10] What had always been a states' rights issue would now potentially fall under federal jurisdiction.

The issue of minors and alcohol escalated after the 1999 Columbine High School massacre in Littleton, Colorado. In the emotional moments after the shooting, legislators pushed for gun-control laws and introduced Senator Hatch's Juvenile Justice Bill. Senator Robert Byrd (D-WV) hijacked the bill by attaching an amendment that allowed state attorneys general to go to federal court to enforce state alcohol laws. David Dickerson, spokesperson for the Wine and Spirits Whole-salers of America, backed the new restrictions but added the caveat that while ordering online was not illegal, "when they [wineries] try to fulfill that order through illegal means such as direct shipment to a customer's home" they should be punished.[11] Senator Thurmond upped the stakes by calling for a probe into whether the Wine Institute had manipulated government officials to craft favorable guidelines for the industry.[12]

Many joined the political opposition to the new bill. Representative Mike Thompson (D-CA) called the Byrd amendment "a measure to protect the business interests of powerful wholesalers who hold many state lawmakers in their thrall."[13] Senator Dianne Feinstein (D-CA) and California governor Pete Wilson wrote to Hatch, arguing against his direct-shipment stance.[14] In the Senate version of the bill, Feinstein sponsored an amendment requiring revocation of winery licenses if illegal sales were made. Congressman Richard W. Pombo (R-CA) argued, "Teenagers are not going to order a twenty-five dollar bottle of wine over the internet and then patiently wait for its arrival in the mail."[15] Robert Koch, head of the Washington, DC, Wine Institute office, argued, "The whole underage access complaint is being driven by wholesalers."[16] De Luca in his well-established style began working on a Wine Industry Code for Direct Shipping.[17]

Alcohol wholesalers now moved more aggressively to lobby Congress for the passage of the 21st Amendment Enforcement Act (HR 2031). The bill placed restrictions on e-commerce, but, more important, it gave state attorneys general the power to use federal courts to enforce state alcohol laws. With the help of Joe Scarborough (R-FL), the 21st Amendment Enforcement Act passed the House of Representatives by a vote of 310–122 and cleared the Senate with a vote of 81–17.[18] The wine industry now faced thirty-eight states with alcohol policies that punished wineries, consumers, and distributors with varying degrees of fines and imprisonment for direct buying of California wines.

Tom Shelton, president of Joseph Phelps Vineyards, believed that "about one-tenth of his business each year is done with customers who place direct orders from roughly a dozen states." He believed that he could triple his sales if the entire nation were opened to direct marketing of wine.[19] With a legislative victory in hand, neoprohibitionists increasingly pressured the twelve reciprocal wine-friendly states for more restrictive laws.

The new electronic commerce restrictions frightened internet companies, and they immediately began pressuring Congress for legislation that favored e-commerce. In a short time, they formed a coalition with the wine industry. The issue shifted from wholesalers versus the wineries to wholesalers versus e-commerce and the wine industry. Through a compromise brokered by Senator Feinstein and Senator Hatch, under the Rules of Construction, an agreement was reached. The new law would have no criminal or civil penalties and provided only for "cease and desist" injunctive relief without any penalties. If you disobeyed the law a second time, there could be additional prosecution. For any state attorney general to use the new legislation, he or she would have to prove that it didn't violate other sections of the Constitution.[20]

SUPREME COURT BOUND

The legal situation worsened when eight states (Florida, Georgia, Indiana, Kentucky, North Carolina, Oklahoma, Tennessee, and pending legislation in Texas) made it a felony to order out-of-state wine. Needless to say, this infuriated many wine drinkers and resulted in the creation of numerous resistance groups. As the protest grew, the California Napa–based consumer group "Free the Grapes" urged consumers and vintners to mail corks to members of Congress to "uncork consumer access to fine wine."[21]

Continued marketing attacks on small wineries resulted in a new series of state laws. Governor George Ryan of Illinois signed into law the Illinois Wine and Spirits Industry Fair Dealing Act of 1999. The law, sponsored by liquor distributors, restricted a winery's right to fire its distributors without "good cause" and allowed the state's Liquor Control Commission to force all parties to continue their relationship. Distributors claimed the law was needed to protect them from foreign suppliers who arbitrarily dumped wines on the American market. The first test of the law came in April 1999, when the Kendall-Jackson winery ended its distribution agreement and, under the conditions of the new law, was ordered to continue doing business with the firm. Jess Jackson responded with a lawsuit.[22] Three years later, a

US district court judge overturned the law on the grounds that it violated the Commerce Clause of the Constitution.

All the chaos over the new felony bills and use of federal courts to enforce state liquor laws propelled De Luca into further action. He was incensed and hurt that after sixty-five years, wine was again being branded as a criminal activity, just as it had been during Prohibition. He asked himself:

> How could they criminalize this activity? This is really something we've been trying to shake loose for sixty-five years. When we were branded as criminals during Prohibition, we had to be fingerprinted. The whole issue of "in bond," the question of paying excise taxes in bond—the originator of that was that this was like skimming profits at gambling casinos. And the moment you moved a product from a part of your winery to another part and prepared for shipment, you'd have to pay a tax on it. All these things all derived from this criminal attitude about our industry. That's why we ended up at the Treasury Department, the Bureau of Alcohol, Tobacco and Firearms, not in agriculture where they have always been in Europe.[23]

John remembered telling members:

> How can we work with partners who use the laws to make it a felony for us to directly sell to a consumer? Criminalization of this activity and the felonization of this activity has got to stop. It's a throwback to the period of Prohibition and it's like they are trying to tar and feather us, you're raising very serious issues about our role in society. By the way if you're not careful, it's going to bring the spotlight of attention to your role. People know what role we play—we produce. People know what role the retailers play and some of them are going to have a hard time understanding their constitutional monopoly in distributing. There are a lot of franchises in America other than the one for wine, beer, and spirits that don't have this state guarantee, and so people are going to start raising questions. Sixty years later, did the 21st Amendment really grant you these kinds of powers to suppress economic competition?[24]

In 1999 the Wine Institute responded to the direct-shipping issues with the publication of its voluntary "Wine Industry Code for Direct Shipping."[25] With the paper, De Luca achieved industry-wide consensus through endorsing the principles suggested by the Coalition for Free Trade, Family Winemakers of California, and American Vintners Association (later WineAmerica), and the recommendations suggested by the National Conference of State Legislatures. The paper addressed the major issues by promoting sales verified with an adult signature in states or counties that permitted alcohol sales. Despite the large array of neoprohibitionist attacks on wine, the California wine industry continued to flourish.

By 2002 the federal courts entered the fracas over direct-shipping bans. Over the preceding years, courts in Virginia and North Carolina had overturned direct-shipping bans because judges ruled that the states had designed their laws to protect their own vintners and, in doing so, had violated the Constitution's Commerce Clause. As expected, the two states appealed the decision. In the meantime, wine-industry leaders anxiously waited for news on direct-shipping bans in New York and Texas.[26] In 2000 Texas judge Melinda Harmon declared that the states' direct-shipping ban was illegal because it violated the United States Constitution's free-trade guarantees. In an attempt to override the federal court decision, the Texas Legislature passed a law in 2001 allowing Texans to ship wine from Texas wineries to their own homes but not wine from out of state. Lone Star wine lovers challenged the law in 2002, and a Third District federal judge overturned the law.[27]

In a direct attack on small wineries, California state assemblyman Marco Firebaugh introduced Assembly Bill 1922 in an attempt to eliminate the so-called gray or parallel market that allowed small wine importers to import and resell wine. The proposal aimed to set a California precedent for the direct-marketing battle. Proponents of the bill fought to shore up the traditional system that designated official US importers who could then sell to retailers. Small-scale importer Michael Opdahl, of Joshua Tree Imports of Pasadena, believed, "This bill is designed to do one thing: consolidate more of the power that the large distributors already have and drive away competitors." Those opposed to the bill feared that it was anti-consumer and would make rare and small-production wines harder to purchase. In response, more than fifty retailers joined with the Family Winemakers of California to oppose the bill and warned that retailers would lose $100 millioin–$200 million in sales and that the state would lose $5 million–$10 million in tax revenue.[28] After a one-month political firestorm, the bill was withdrawn from the docket. This did little to resolve the mass confusion over direct-buying issues, and the federal courts continued to slowly resolve the problem state by state.

In 2003 a Florida federal court ruled that the state's ban on interstate shipments was unconstitutional, but Judge Richard Berman stayed his decision, knowing that his ruling would be challenged. A short time later, a federal court in Michigan ruled that their state's wine-shipping bans were illegal.[29]

Both sides of the direct-shipping controversy began to expect that the final decision would lie with the United States Supreme Court. Planning for this outcome, the Coalition for Free Trade hired lawyer Kenneth Starr, of Whitewater and Monica Lewinsky fame, to prepare for the eventual High Court case. Starr believed that these laws were "designed to handicap and hinder out-of-state wineries" and therefore violated the Constitution.[30] The preparation proved prudent, and in 2004 the Supreme Court agreed to hear the Michigan and New York direct-to-consumer appeals. The highest court in the land would have to decide the outcome of the conflict between the Twenty-First Amendment, which assigned the power to control alcohol sales and distribution to states, and the Commerce Clause, which allowed free trade between states.[31] The unified wine industry waited for the decision, and most realized that major labels would still be handled by wholesalers and hoped that small wineries would win a chance to compete for their market share.[32] Stephen Evans, BBC North America business correspondent, summed up the collective angst in this way:

> It is hard to know what the court will decide. The case has gone through a string of lower courts, each swerving different ways. Three of the Supreme Court justices have in the past indicated that the 21st Amendment over-rode the right to trade freely. Two others have suggested the opposite. It is part of the paradox of America. This country is devoted to the free market but with strong nanny-ish tendency, especially where matters of drink (and sex) are concerned. The people should be free—as long, that is, as they don't buy a delicate chardonnay directly from its maker.[33]

As the cases wound their way through the Supreme Court, state attorneys general continued to enforce their own alcohol laws. Massachusetts state attorney general Tom Riley, with the assistance of the Massachusetts Alcohol Beverages Control Commission, set up a sting operation to shine a light on direct-sales infractions. As part of the operation, five underage volunteers made eight online purchases of alcohol using their own credit cards and addresses. Riley then issued arrest warrants for ten retailers and UPS, DHL, and FedEx for not verifying the ages of the buyers. The Wine and Spirits Wholesalers of America said the operation proved the need for the three-tier system.[34]

In 2005 the United States Supreme Court ruled five to four in Granholm v. Heald, case 544 US 460, against Michigan and New York. The justices held that because the direct-shipment laws had discriminated against out-of-state wineries, the state laws violated the US Constitution's Commerce Clause.[35] Justices Anthony Kennedy, Antonin Scalia, David Souter, Ruth Bader Ginsburg, and Stephen Breyer struck down the Michigan and New York laws. Justice Kennedy wrote, "States have broad power to regulate liquor under the 21st Amendment, but this power, however, does not allow states to ban, or severely limit, the direct shipping of wine while simultaneously authorizing direct shipment by in-state producers."[36] Their decision clearly allowed direct shipment of wine as long as in-state and out-of-state wineries were treated equally.

Unfortunately, the decision did not set a clear path for the future of national direct sales. The states of Florida, Connecticut, Indiana, Massachusetts, Ohio, and Vermont still had discriminatory laws on the books, and state legislatures had to create new policies and laws. Thirteen states, including California, had "reciprocity" laws, permitting wine shipments from other states that allowed out-of-state wineries to ship to their residents. Many believed this system discriminated against nonreciprocal states. Some states considered laws requiring wineries to register and pay taxes in order to sell within their state. Complicating the mix were the fifteen states that banned wine shipments altogether.[37] In a way, Prohibition still ruled the day, and a cohesive wine direct-buying system would have to be resolved at a later time.

{ 11 }

Politics of Wine

Bipartisanship, Sound Bites, and the Wine Caucus

> The philosophy I have is that the Wine Institute serves as the bridge between the urban left and the rural right, that the Wine Institute's attitude is there is no Republican or Democratic way to drink wine, and that we have—between those who produce the product and those who consume the product—a natural alliance between the countryside and the cityside. Our consumers are primarily in the major cities, and the producers are in the country.
>
> —JOHN DE LUCA, "President and CEO of the Wine Institute," oral history

De Luca's dream of an American wine culture required more than marketing skills. His leadership role at the Wine Institute necessitated that he recognize and address the political processes that could produce favorable state and federal governmental policies. This suited De Luca, the political scientist, politician, and strategist, and made him the perfect candidate for the battle against neoprohibition. After two decades, John realized that the best gift he could give the institute would be a place at the table in Washington, DC, and at statehouses across the nation. Throughout his years at the helm of the Wine Institute, he laid the groundwork to make the wine industry a continuing topic for government officials and the media.

BIPARTISANSHIP

From the start, De Luca recognized that the wine industry, by virtue of its rural agricultural members, was predominantly Republican and that most members were conservative on economic, social, and policy issues. De Luca also understood that in the world of politics, one party is not always in the majority and that the industry needed to reach across the aisle to both parties. He knew he could not change the political values of his membership, so he shifted the Wine Institute's political

135

connections and contributions to reflect a more bipartisan approach. He began by creating a political action committee that promoted balancing contributions to both Democrats and Republicans. As proof he offered, "Over a period of twenty years we ended up, like most recently, where we supported Dianne Feinstein [Democrat] strongly but we also supported Pete Wilson [Republican] strongly, and so—no contradiction in that."[1] No matter which party controlled Congress, the executive branch, governorships, or state legislative branches, the Wine Institute sought politicians who were favorable to the industry. Throughout the 1980s and 1990s, De Luca felt that this tactic worked.

> The benefit of this bipartisanship is apparent as when we just went back to Washington, which is now in Republican hands, we had a wonderful reception. Congressman Bill Archer is Republican—chairman of the Ways and Means Committee. Congressman Bob Livingston of Louisiana is Republican chairman of the Appropriations Committee, and the majority leader, Bob Dole. We had a very good balance of Republicans and Democrats, as we do in this state. In fact, I just received a letter from Cal Dooley, who was Congressman Dooley from the Valley, who said on the Market Promotion Program (MPF) vote, of the 268 votes that we had, there were 132 Democrats and 136 Republicans. Can you believe anything more bipartisan than that?[2]

The only problem with bipartisanism was the fact that some candidates whom the institute supported ended up losing to candidates without proven track records on wine issues. In one case, Wine Institute member and Republican George Radanovich won California's Nineteenth District in 1994 without the endorsement of the Wine Institute. The membership had endorsed his opponent, incumbent Democrat Rick Lehman, who in the past had supported the Wine Institute and its agenda more robustly. De Luca remembered:

> Here's a person who had been working with us, had been very assertive on our behalf, and a member of the Wine Institute running against him, whom we did not support—as an organization—but individual members and wineries did. The attitude there was, we're not supporting Democrats or Republicans; we are supporting those

who helped us, Republican or Democrat. Even though Congressman Lehman lost, it was a message and a signal to everybody that we do stand by our word, and that if you work with the Wine Institute, we will not abandon you just because one of our own members is running against you. What better test could you have, its given us enormous credibility.[3]

Luckily, in this case, all turned out well, as Radanovich went on to become a cofounder of the Congressional Wine Caucus (CWC).

POLITICAL SOUND BITES

Drawing from life experiences and academic training, John honed his discourse skills and learned the necessity of occupying the linguistic high ground. In politics he firmly believed that the person

who defines issues, wins out in elections. You spend billions of dollars defining your opponent, and what kind of sound bites do you have? "Build a bridge to the 21st century?" "Law and order;" all the different terms that you know that are hammered away. The commanding heights of language govern. And look what just happened with regard to our recall in California [the 2003 recall of California governor Gray Davis]. What language, what were the sound bites that were used? Bustamante lost with his "tough love," but Schwarzenegger won with his, "I got a broom, and I'm going to sweep out the miscreants over there in Sacramento."[4]

During De Luca's struggles to rebuild an American wine culture, he used language that played a large role in determining the fate of wine. Like a teacher in his university classroom, he drove the key points home by repeating key phases and words that then became the language used by scientists, government officials, and the media to define the issue. He recalled:

So I have been a student of language, and therefore we had to get back to our fundamentals, which was not to overstate, not to really just be organaleptic, which means aroma and bouquet, which is wonderful for one part of our world. But in this field, we have to be seen as part of nutrition, fighting disease, supporting the world of science. . . . As a teacher, which I am, you use repetition, but not always word for word. The staff has understood it. It's been part of my mentoring to my staff.[5]

Most important was the transference of the defining language to the media, as they described the Wine Institute talking points to the general public. "It's been what I have preached to the press. It's been to get those who cover us, and those who speak for us, and those who are the policymakers the thrust of our new policies, e.g., getting the dietary guidelines changed in 1995 so that the headline in *The New York Times* read, 'Wine Good For Your Heart: Federal Turnaround.'"[6]

For De Luca, the currency of politics was language. Therefore, he was careful to craft his talks, speeches, white papers, news releases, and membership reports to meet the level of the policy direction needed. For him there was "no question about it, how you're called, how you define an issue, the definition of terms is so important. And I felt, coming into the industry at that time, that the words like *sin tax, gateway drug, drug of choice, street drug,* that that whole amalgam of words was making incredible inroads into the image and into our standing."[7] From the start, De Luca stayed on message and defined winemaking as part of the nation's agricultural heritage. He recited the fact to anyone who would listen—that winemakers were farmers and always referred to them as "wine-growers" who farmed "wine-farms." He knew that the Jeffersonian image of the independent and patriotic farmer played well in conservative, rural areas and that it could move many naysayers beyond the neoprohibitionist barrage of negative nicknames.

CONGRESSIONAL WINE CAUCUS

De Luca realized that he needed a coalition of key congressional organizers supportive of wine causes, and two names rose to the top. First was George Radanovich (R-CA) of the Sierra Nevada Foothills Radanovich Winery, who entered the House of Representatives in 1995, representing California's Nineteenth District (Fresno County). He was joined in 1998 by incoming freshman member Mike Thompson (D-CA), a wine-grape grower and state politician from California's Fifth District (parts of wine country in Contra Costa, Lake, Solano, and Sonoma Counties). With the assistance of the Wine Institute, the two representatives hosted wine-tasting receptions, vintner "meet and greets," and dinners to meet fellow congresspeople and to solicit their assistance. Their goal was to create a bipartisan and bicameral Wine Caucus.[8]

In a short time, they had fifty-three representatives and five senators in the caucus. Congresspeople from California, New York, Tennessee, and North Carolina were joined by Senators John Warner (R-VA) and John Breaux (D-LA), as well as by House Minority Leader Richard Gephardt (D-MO). Their first task was to battle Senator Strom Thurmond and Orrin Hatch on their attempts to restrict wine distribution and sales. As we have seen, this proved to be an uphill battle, as Thurmond crusaded to raise the excise tax on wine, to block the use of directional labels, and, with Hatch, to promote internet restrictions. At the same time, caucus members also played a key role in securing funds for research on wine health issues. To keep the group unified, De Luca hosted events at the Library of Congress and adopted this strategy: "It is all about wine and socializing. No speeches."[9]

De Luca and the Wine Institute staff considered good relations with members of the Wine Caucus as a top priority. Reaching back to his ethnic and family roots, John always made sure to incorporate the idea that the wine industry was a family agricultural endeavor. One year, as part of Wine Appreciation Week, John had arranged for vintner visits to the White House. Bruce and Paula Rector, of Glenn Ellen Vineyards & Winery, asked John if they could bring their sons, aged ten and twelve. This excited De Luca, who thought it would be a historic opportunity for the boys. So John helped the Rectors and other vintners bring their children. He remembered: "They were there when we had the meeting with the president, and they, along with the children from other vintners who came, permitted us to point out that far from high-powered attorneys and lobbyists, that here were families. They introduced themselves and went into the fact that they were first-, second-, and third-generation in terms of their families. So that really was, I think, visually, very important."[10]

To help solidify the idea that wine was an agricultural product, De Luca sometimes brought state legislators and congresspeople to the vineyards to help them better understand the process of wine making. For De Luca, these visits were important.

It's not just a joyride to come out here and just drink wine and talk, though that's always a wonderful attribute of coming out here. We're very serious about educating, through exposure, through trips. Nothing as instructive as coming out here and literally walking our vineyards. You find out about our agriculture, you find out about the natural components, that we're not manufacturers as much as we are really farmers, and the soil and the stewardship of the soil, and the issues that we're concerned about, like pests, like phylloxera, like Pierce's disease, like Eutypa.[11]

As the direct-shipping controversy rocked the industry, these visits gave politicians the time to debate with winemakers and consumers about the problems surrounding access to certain wines. Concerns over direct shipping intensified when national media outlets, such as *The Wall Street Journal, The New York Times, USA Today,* Knight Ridder, the *San Francisco Chronicle,* and the *Los Angeles Times,* featured editorials and journalistic articles that promoted the pro-wine message. As a result of the articles, politicians in Washington, DC, received numerous complaints from constituents who could not get their favorite wines. Media coverage and consumer concerns strengthened the cause, but it would be members of the Congressional Wine Caucus who provided the pressure to remake policies.

A CONTINUED PLACE IN AMERICAN POLITICS

At the time of the 117th Congress (in 2020), seventeen years after De Luca had left the Wine Institute, the Congressional Wine Caucus had 187 bipartisan and bicameral members. Congressman Mike Thompson remained as cochairman and was joined by Congressman Dan Newhouse (R-WA, Fourth District), an agricultural scientist, who replaced George Radanovich. As always, they had the stated organizational goal of working together "to protect America's viticultural heritage and strengthen our vibrant wine community from grape to glass."

The caucus now has its own Web presence and a posted introductory statement that reads: "Members of the CWC are joined not only by their deep appreciation of wine, but also their understanding of the wine community's significant economic impact. Wine is produced in all 50 states and generates jobs and revenue in both rural and urban areas. The wine community contributes an estimated $220 billion to the U.S. economy annually and creates the equivalent of more than one million full-time jobs."[12] The caucus webpage also lays out their legislative and policy goals:

- Passing legislation to conserve and protect vineyards and open spaces
- Advocating for viticulture research and management funding
- Combating European Grapevine Moth and the Glassy-winged Sharpshooter
- Fighting for fair market access for our wines in foreign markets
- Protecting wine's access to the Customs Duty Drawback program
- Enacting and extending bipartisan legislation to modernize the Wine Excise Tax
- Recognizing the value and contributions of American wines and grapegrowing regions[13]

Missing from the goals are some of the hardest-fought policy, legislative, and scientific battles led by De Luca. There is no mention of drinking in moderation, the health benefits of drinking wine, or the rebuilding of an American wine culture. Had neoprohibitionists finally given up the battle, or was this just a lull in what seemed to be an eternal struggle? De Luca believed that he had left the Wine Institute with the policies and tools needed to continue pushing back against the anti-alcohol effort, but he also warned that a prudent wine industry must always maintain a vigilant guard to protect wine from anti-alcohol forces.

Establishing the Path for
the Wine Institute's Future

After his retirement in 2003, De Luca set out to institutionalize many of the important alliances that he had created or joined. One example was the Agriculture High-Tech Alliance that had played a major role in fighting Senator Orrin Hatch's 21st Amendment Enforcement Act. In the bill, wholesalers attempted to shift state internet and alcohol legal cases to federal courts. Many in the high-tech community had worried that any sort of federal participation or regulation would affect all internet sales. As a result, they had found common cause with the Wine Institute, and with De Luca's assistance they had allied with the Farm Bureau and the Western Farm Growers. As long as De Luca was around, the alliance had lasting importance. John feared that in his absence, the alliance would fall into disrepair: "I was at the semiconductor industry annual awards dinner with all of the high tech CEOs, and mingling with them, and Governor George Pataki from New York was there, who is a dear friend, with his wife Libby. Well, absent continuing those relations, they sort of died on the vine."[1]

His efforts to reach to old allies who had fought previous attacks on the wine industry continued. De Luca remembered working with the California Teachers Association as deputy mayor and the assistance they gave the institute during the California excise-tax initiatives. Because of this relationship, John reciprocated by serving as a trustee on the CTA Institute for Teaching board. After three decades of working with the teachers union, he felt that the alliance needed to continue. He also formed other alliances by serving as a member of the World Affairs Council, as senior adviser to the president of the University of California, serving as an advisory board member for the president of Cañada College, and as a member of the Commission on Agriculture and Natural Resources for the University of California; by visiting as a distinguished scholar at UC, Berkeley; by teaching at the UC, Berkeley,

Goldman School of Public Policy; and by working at the Harvard John F. Kennedy School of Government. In 2007 the Alumni Associations of the University of California named John AAUC Advocate of the Year for his work "advancing the value and critical importance of research to California's future, building stronger relationships between UC and the state's agricultural industry."[2]

One of John's favorite endeavors included being chairman of the board for the University of California, San Francisco, Ernest Gallo Clinic and Research Center. Gallo had founded the center in 1980 as a nonprofit, multidisciplinary research institution devoted to the study of neuroscience. Originally, the center had funded research on molecular, cellular, and neuronal mechanisms that underlie alcoholism and substance abuse. In more recent times, the United States Army granted the center $15 million to "accelerate the discovery and development of new medications to treat alcohol and substance abuse in the context of post-traumatic stress and combat injury." After being renamed the Institute for Molecular Neuroscience, the program brought together "a team of national experts who were unaffiliated with grant applicants to conduct an independent, peer review process."[3] Many of their discoveries are now utilized to treat war veterans, athletes, and civilians suffering from traumatic head injuries.

Keeping these kinds of alliances alive became a primary postretirement goal for De Luca. He believed that the "position of executive vice chairman permits me to still stay in the industry, to not have day-to-day operational responsibilities, but to stay in liaison, advisory; special projects relationship to the industry. And I think that this is extremely valuable for the Institute, and for my wife and for me."[4] Members of the institute also recognized the value of the associations, and as a result De Luca signed two additional five-year contracts.

As his semiretirement progressed, De Luca reflected upon his quest to restore America's lost wine culture, and he formalized a historical perspective of how anti-alcohol beliefs affected American wine history:

> I have a great thesis that I have shared with many people, and that was that the turning point in this great landmass that the geographers used to call terra incognito, the cartographers used to say "uncharted land." It occurred during a period of what we call the French and Indian Wars; Queen Anne's War, King George's War, King William's War. The French and the Algonquins were winning.

The British couldn't win that war with the French unless they changed their tactics, brought in new resources, brought in new training methods for their army. The British, and not the French, won that big battle over who was to command the Eastern seaboard during that period of the 1720s to the 1760s. If the French had won, we would have had French literature, we would have had French cuisine, we would have had French culture.

If the British settlers had been the London merchants, who were very cosmopolitan, and went to Madera, and went to Spain, and went to Oporto, if it was that part of England that settled the United States, there would have been an entirely different culture in the United States from the British side. We would have had great institutions of parliament, we still would have been able to read Shakespeare in the original English, but we wouldn't have had the most pinched culture then existing in Europe, which was the Puritan one, that was so self-negating that the British couldn't stand them and they couldn't stand the British, so they left. As a result the people that we venerate, who came in the early days with the Mayflower, the founding Fathers and the great tradition of the first Thanksgiving, those people were very brave and very adventurous and had very strong ideas, but their culture was very narrow.

So what got established in this new land with the victory of the British over the French was that set of principles that they had, this Puritan ethic, which then migrated to the Bible Belt. So in terms of what happened to this land, where the French did settle, like New Orleans and Montreal, you have an idea of what it would have been like had Montcalm [Louis-Joseph, Marquis de Montcalm] and not Wolfe [James Wolfe] had won the Battle of the Plains.

So all the Italians and Yugoslavs and Greeks and French and Spaniards and Latin Americans who came in the nineteenth and twentieth century, had to adapt to the existing one. The Germans who came, of course, brought a beer culture which was terrific for America because of the national sport being baseball, and breweries wanting a cold drink in those hot, humid days when Americans were playing baseball. They had a brewery in every little town. You didn't have a winery in every little town. And the pioneers who came across the United States in the Conestoga wagons, they are the ones who initiated the Pennsylvania Whisky Rebellion, but they were not drinking Sauvignon Blanc and they weren't drinking Merlot.

But they did settle out there in California. So we really had one part of the country, primarily in Northern California, and waves of immigrants who came, who did influence how Prohibition was executed, because if you remember, under the Volstead Act, there were exemptions to the complete ban of the production of any intoxicating liquor. Wine was never mentioned during Prohibition. It was in the definition of terms of what was an intoxicating liquor, and that's how wine was drawn into it, even though many people in California at first didn't feel that they were affected by the Volstead Act and the various state acts and then the 18th Amendment. But there were exceptions based on that culture, and that was that you could have sacramental wine out of deference not only to Catholicism and to Anglican influence, but also to Judaism, that you could have wine for home consumption. That is why you had all during Prohibition the growers exporting grapes all throughout the United States with trains, with big immigrant populations waiting for the grapes to come so that they could make wine in Chicago and Pittsburgh and Philadelphia and Boston and New York. My uncles tell me of waiting for the trains from California.

All of that really derives from cultural underpinnings, and so even though we look at our laws as though they were preordained and static, it really has in fact been the way America was settled, the way the different states looked at such basic things as the consumption of alcohol. There was a big difference in the big cities versus the rural part, and that became part of Roosevelt's coalition, the wets and the dries. In fact, it's very instructive to think that Roosevelt and the Democrats in the 1932 Democratic convention had as their plank the end of Prohibition, and that part of their coalition, the Rooseveltian coalition, included not only academics and the Jewish community, but all of these teeming, drinking Irish, Italian, Polish, Greek, and Yugoslav citizens.

When we look at the history of the United States, what could you do today to acknowledge the past? How do you compensate for the fact that we did not have an evolution in the United States that would have been more East European, West European, Mediterranean, based on the number of people who came later? How do you address today's society to give yourself a chance to make it part of the lifestyle and culture of your country? You can't pass a law-on one hand, Prohibition, and not expect many people to disregard the law and go to speakeasies. In effect, we subsidized the criminal elements of the United States and underwrote the strengthening of the Mafia in America with Prohibition. So laws by themselves didn't work.[5]

John used this historical narrative to frame the intentions of his adversaries. The puritanical part of America fought the perceived evils of alcohol, while wide swaths of later immigrants embraced beer and wine cultures. This led to a centuries-long, American, on-again, off-again national struggle with all alcoholic beverages.

In the latter half of the twentieth century, De Luca became the face of a national rebuilding of an American wine culture. At times he felt he was losing the battle, as when the Federal Trade Commission informed him that heart-health advertising would not be tolerated unless wine could be investigated by the FDA like all other foods and drugs. This meant that wineries would have to go through the FDA approval process and be put through rigorous investigations just like such drugs as Verapamil, Lanoxin, and Allopurinol. He realized that in Shakespearean terms, "therein lies the rub," and that he had to continue the fight. Instinctually, scientifically, and politically, he realized that the only way to win this argument was through third-party scientific studies. Thus, he shifted the Wine Institute mission statement from marketing and public relations to policy initiatives intended to repair the good name of wine.

THE FUTURE OF WINE AND THE NEVER-ENDING NEOPROHIBITIONIST BATTLE

De Luca fully believed that the battle with neoprohibitionist forces was never-ending and that members of the wine industry must understand that "the price of liberty is eternal vigilance, and that each generation can find a different approach to larger questions of stewardship, of sustainability, of social responsibility."[6] He warned that in the long run, only corporate responsibility, family ethics, and name-brand pride would allow the industry to prosper.

> There is a family connection here; we do put our names on our bottles. That is very unique, that people actually say, "You can judge me." Well, very early I had Jack Valenti, as I had a lot of other people, come to the board of directors to address them on bigger issues. Jack was a leader in the Motion Picture Industry, and Louis Martini was the chairman of the board that year, so that was 1976, 1977. Jack spent some time with him, and his first words as he was addressing the board of directors were, "I had never met a label before." And of course, you know, in an assembly-line society, who knows who made your ready-made suit or your shoes or any of your products? But here, people put their name on the label.[7]

John had spent his entire Wine Institute career carrying the banner for the effort to build credibility for the wine industry and to reestablish an American wine culture. He worried that a loss of historic memory could wipe away all the changes that he had fought for.

Epilogue

Who Will Carry the Banner Next?

Throughout the twentieth century (and even more in recent times) a loud and politically active sector of the American populace fought first to prohibit alcoholic beverages and then later worked to reduce alcohol consumption by regulating distribution and increasing taxes. Mainly centered in the rural South and central states, this voting minority drew motivation from puritanical and evangelical beliefs intended to save the nation from the perceived moral degradation of the evils of alcohol. They labeled alcohol supporters as sinners, degenerates, morally weak, and un-American, and they blamed a great number of societal ills on demon rum. Their first major victory came with the Prohibition Amendment that successfully waged war on large alcohol-related corporations. As a result, Prohibition moderated American alcoholic consumption, ended an era of working-class saloons, served as one of the more successful alliances of the upper and middle classes to legislate morals and habits, and increased consumption of high-alcohol sweet wines.

To accomplish their political victory, dry forces had learned to utilize politics that included shaming, equating temperance with a godly cause, grassroots electioneering that threatened political careers, and ignoring scientific discoveries and scientists, and they learned to form alliances to build their political base. Sometimes they even embraced their enemies in what has been referred to as "Bootleggers and Baptists" coalitions. Anti-alcohol, aggressive, grassroots political strategies proved successful, and in a short time America lost a wine culture based on the European tradition of wine as a moderating mealtime beverage.

What anti-alcohol zealots forgot was that urban centers, teeming with immigrants from alcohol-consuming cultures, had populations that exponentially grew to outnumber rural voters. As a result, urban voters took back control of federal alcohol policies and ushered in a negotiated Repeal Amendment. But after more than a decade of controlling alcohol policies, dry forces were unwilling to totally relinquish their alcohol jeremiad. After their stunning defeat, they kept their cause alive by negotiating a states' rights compromise that allowed individual states and local governments to prohibit, regulate, and tax alcohol. The ideological struggle was simply transferred from the federal government to state and local governments.

Over the last eight decades of the twentieth century, a few notable individuals carried the banner for the wine-moderation cause. To that end, men like Leon Adams helped establish the Wine Institute as a means to reeducate Americans and rebuild the intellectual, moral, and business structure of the commercial wine industry. The institute's initial successes in the decades of the 1930s through the 1960s slowly reestablished American's interest in table wines. But their victories began to falter by the 1970s, as neoprohibitionists began a new quest to limit distribution and sales of alcohol through increased regulations and heavy taxes. This is the Wine Institute that John A. De Luca inherited. As a result, he spent his entire wine-industry career battling the fearmongering of anti-alcohol forces, who portrayed wine as an illicit drug requiring restrictive policies and laws. During his tenure at the institute, De Luca drew heavily from his religious, family, ethnic, political, scientific, and academic values to help set in motion government policies favorable to reestablishing an American wine culture. Like the first urban voters who overcame rural advocates of Prohibition, De Luca drew much of his energy from his Eurocentric background and beliefs.

Upon his retirement, De Luca left the helm of a thriving industry with strong hopes for increased domestic and foreign sales. Yet he warned members of the Wine Institute that the battle against anti-alcohol forces, like the continuous threats from vineyard pests and diseases, was far from over and that the struggle would be never-ending. He reminded enthusiasts and wine producers that over the past century, dry ideals had repeatedly threatened the wine industry and that they must maintain constant vigilance in order to combat prohibitive state and federal policies.

De Luca understood that every time the industry became perceived as just another manufacturer of alcohol, they suffered a resurgence of neoprohibitionist anger. He understood that parts of society did not believe in the consumption of alcohol because of religious beliefs, health concerns, or worries about being prone to alcoholism and that they would never give up their anti-alcohol campaign. He warned winemakers not to forget the lessons of Prohibition and of overregulation brought on by the kind of corporate greed that eclipsed social and moral responsibility. To demonstrate his argument, De Luca vigorously reminded members of cautionary tales recalling how the California Wine Association's 1890s monopoly fell to Prohibition and how the consolidation of the industry during the Whiskey Revolution of the 1950s through the 1970s rekindled neoprohibitionist attacks from the 1980s to the 2010s. Time and again, De Luca also reminded grape growers, winemakers, suppliers, and consumers to never forget the ancient tradition of drinking wine in moderation for meals and health. In his words:

> I've told our people that there is no question that it applies to us, you know, that the price of liberty is eternal vigilance, and that each generation can find a different approach to larger questions of stewardship, of sustainability, of social responsibility. We happen to be put in a very unique position as the only industry that I know of in America where we had the 18th and the 21st Amendments to regulate our industry, therefore we have a very special responsibility, and so the terms "corporate responsibility," "social responsibility," and "families," are important.[1]

As De Luca retired from his everyday duties at the Wine Institute, he continued to speak about maintaining vigilance:

> Certainly don't take anything for granted in our society, how quickly things can change. We recognize watching the politics and the political debate in our country how overnight people can come out of nowhere and be suddenly thrust in the limelight, or suddenly some things that we thought were eternal are not. So given the nature of communication, given the nature of activism, given the nature of just public debate, the wine community should not forget its last hundred year history. Cannot. Not to be in any way cynical, but to be what I would consider being healthy skeptics. I think there's a

big difference, this industry has to be healthy skeptics about what happens in our society, but that the burden is still on us.[2]

Over the past decade, as De Luca's presence and influence waned, the wine industry continued to expand, and California wine became a major player in the global marketplace. Large corporate wineries profited, and smaller estate wineries became destinations for tourists and wine enthusiasts. Despite the profitable growth, a few remembered the warnings of the De Luca era and worried about a resurgence of neo-prohibition. Cyril Penn, a *Wine Business Monthly* editor, called for a return to the De Luca days of championing moderate alcohol consumption and health discussions. He believed that in the time since De Luca's retirement, the Wine Institute had abandoned the dissemination of scientific information on wine health and moderation and returned to being just a marketing arm of the industry. Lewis Perdue, an entrepreneur, technologist, scientist, author, publisher, and journalist, responded that "the Wine Institute's abandonment of the issue has squandered that progress and his [De Luca's] legacy and created a vacuum that offered a clear road to Neoprohibition." Perdue believes that De Luca "seized the public notion that wine held a certain high ground among alcoholic beverages, and made substantial public and regulatory progress toward separating moderate consumption from abuse." Both Penn and Perdue believe that De Luca "pressed for a middle ground of providing credible, third-party information without trying to obscure access to those with an anti-alcohol viewpoint."[3]

Their warnings and call for a return to De Luca–style education, scientific research, and political activism seem to have fallen upon sterile soil just at a time when new religious and scientific attacks again threaten the industry. Conservative evangelical ministers continue to preach about the perceived evils of alcohol. In a 2018 *ChurchLeaders* magazine article, Barry Cameron, senior pastor of the Texas Crossroads Christian Church, lamented the apparent downfall of Olympic gold-medal snowboarder Shaun White after his arrest for vandalism and public intoxication. In the article Cameron asked the central question "Can a Christian drink alcohol?" and answered that it does not enhance the drinker's testimony. He continued:

For a minute, forget about making a definitive case for or against "drinking" from the Bible. Here's the truth from logic and real life. No one starts out to be an alcoholic. Everyone begins with a defensive attitude saying, "I'm just a social drinker and there is nothing wrong with it!" No one says, "It is my ambition that someday I want to lose my job, my health, my self-respect, my marriage and my family. Someday I want to be dependent on alcohol to get through my day." Yet, this is the destination at which several millions of people have arrived. Why do you suppose that is? It is because alcohol is promoted and elevated as a normal sophisticated activity in life. . . . It is also expensive, addictive, and enslaving. People get hooked by America's number one legal drug. And just like all illegal drugs, alcohol finds its way into the body, the bloodstream, and the brain of the user/abuser.[4]

Recent scientific studies have begun attacking the idea of alcohol moderation. A 2018 *U.S. News and World Report* article touted this headline: "No Amount of Alcohol Is Healthy." The piece referred to another article published in *The Lancet* that focused on a study conducted by the prominent University of Washington's Institute for Health Metrics and Evaluation. The study supported the idea that heavy drinking had increased international rates of disease, disabilities, and death. The authors attacked health claims made by alcohol-moderation proponents and stated, "Consuming zero standard drinks daily minuses the overall risk to health." Researchers wrote that "any supposed boosts to health are massively offset by the costs." *The Independent* news outlet went further and stated, "The myth that one or two drinks a day are good for you is just that—a myth. This study shatters that myth." *New York Times* reporter Nicholas Bakalar reported, "Just one alcoholic drink a day slightly increases an individual's risk for health problems, according to a large new study. A new group of scientists had begun reviving the idea that moderation is the wrong approach and that any amount alcohol is bad for a person's health—a finding that runs contrary to much previous research and public health guidelines in many countries."[5]

Media coverage attacking moderation and alcohol abuse increased. *U.S. News and World Report* staff writer Katelyn Newman warned of the continuing ill affects of alcohol consumption in a February 2020 article titled "Study: Alcohol Is Fueling an 'Urgent Public Health Crisis' in the U.S." Newman reported that Neal Freedman, a senior researcher for the National Cancer Institute, believed that "alcohol is killing Americans across boundaries of age, gender and race." Further rumblings continued, and in 2020 *U.S. News and World Report* writer Gaby Galvin stated that, "Centers for Disease Control and Prevention researchers estimate that an average of more than 93,000 people died from a cause attributable to excessive alcohol every year from 2011 to 2015, at an annual rate of 27.4 deaths per 100,000 people." The article continued, "Little progress has been made in preventing deaths caused by excessive drinking." Additional alcohol-abuse concerns arose, as researchers found that alcohol consumption increased during the 2020 COVID-19 pandemic. The news forced federal researchers preparing for the five-year review of the U.S. Dietary Guidelines to consider rewriting their moderation stance to recommend only one drink a day for both men and women.[6] As the Donald Trump administration shifted the nation further to the conservative, religious Right, many began to wonder when new anti-alcohol forces would again wage war on alcohol. The question remains: Who will carry the banner next?

Notes

ACKNOWLEDGMENTS

1. John De Luca, "President and CEO of the Wine Institute, 1975–2003," oral history.

2. Over the past decade the author, as part of his study of the Italian foodways diaspora, has visited farms, producers, food schools, and markets in Catania, Palermo, Piana degli Albanesi, Vallelunga Pratameno, Cefalù, Perugia, Florence, Modena, Parma, Bologna, Spoleto, Reggio Emilia, Venice, and Rome.

INTRODUCTION

1. Lisa McGirr, *The War on Alcohol: Prohibition and the Rise of the American State,* xvii.

2. Victor W. Geraci, *Wine by Design: Santa Barbara's Quest for Terroir.* The title of this book is a phrase first used by Bruce Sanderson in "Wine by Design," *Wine Spectator,* November 15, 2001.

3. Vintibusiness is the vertical and horizontal organization of wine-grape farming, wine making, distribution, and marketing of wine. Victor W. Geraci, *Salud: The Rise of Santa Barbara's Wine Industry.*

CHAPTER 1. THE EARLY REPUBLIC'S FAILURE
TO ESTABLISH A WINE CULTURE

1. "Wine Truly a Drink of the Ages," *Los Angeles Times,* June 6, 1996; Paula Harris, "Wine Medicinal Nutritional Roots Traced Back to Stone Age," 29–30; Kim Marcus, "New Tests Find Evidence of Wine at the Dawn of Civilization," 8.

2. Carlo Cipolla, "European Connoisseurs and California Wines." Discussions of the chicken or egg argument for the beginning of American agricultural capitalism can be found in Allan Kulikoff, *The Agrarian Origins of American Capitalism;* and James A. Henretta, *The Origins of American Capitalism: Collected Essays.*

3. Tim Unwin, *Wine and the Vine: An Historical Geography of Viticulture and the Wine Trade,* 3.

4. Richard Hakluyt, *The Principal Navigations, Voyages, Traffiques, and Discoveries of the English Nation,* 51–56.

5. Robert Johnson, "Nova Britannia," 16.

6. Michael R. Best, "The Mystery of Vintners." Practical wine-grape growing guides in English existed as early as 1658.

7. *Journal of the Commons House of Assembly, 1742–1744,* ed. J. H. Easterby, in *Colonial Records of South Carolina,* 553.

8. Thomas Pinney, *A History of Wine in America: From the Beginnings to Prohibition,* chap. 3.

9. Leon D. Adams, *The Wines of America,* 17; Pinney, *From the Beginnings to Prohibition,* 84–85.

10. Joni G. McNutt, *In Praise of Wine: An Offering of Hearty Toasts, Quotations, Witticisms, Proverbs, and Poetry Throughout History,* 108.

11. Paul Lukacs, *American Vintage: The Rise of American Wine.*

12. Pinney, *From the Beginnings to Prohibition,* 107–14.

13. Pinney, *From the Beginnings to Prohibition,* 121–25.

14. Pinney, *From the Beginnings to Prohibition,* 139–49.

15. Lukacs, *American Vintage,* 15.

16. Lukacs, *American Vintage,* 35.

17. William J. Rorabaugh, *The Alcoholic Republic: An American Tradition,* 105.

18. Lukacs, *American Vintage,* 35.

19. Don Kladstrup and Petie Kladstrup, with J. Kim Munholland, *Wine and War: The French, the Nazis, and the Battle for France's Greatest Treasure,* 240.

20. Ian R. Tyrrell, *Sobering Up: From Temperance to Prohibition in Antebellum America, 1800–1860,* 147.

21. Tyrrell, *Sobering Up,* 11.

22. Rorabaugh, *Alcoholic Republic,* ix.

23. Rorabaugh, *Alcoholic Republic,* ix.

24. Rorabaugh, *Alcoholic Republic,* 61–89.

25. Tyrrell, *Sobering Up,* 232–33.

26. Stanton Peele, "The Conflict Between Public Health Goals and the Temperance Mentality," 805-10.

27. Peele, "The Conflict Between Public Health Goals and the Temperance Mentality."

28. Peele, "The Conflict Between Public Health Goals and the Temperance Mentality."

29. Peele, "The Conflict Between Public Health Goals and the Temperance Mentality," 104.

CHAPTER 2. THE RISE OF THE FIRST AMERICAN WINE CULTURE

1. M. K. Bennett, "Climate and Agriculture in California," 158–63.

2. James J. Parsons, "Uniqueness of California," 45.

3. *Gemeinschaft* and *Gesellschaft* (generally translated as "community and society") are categories coined in the late nineteenth century by German sociologist Ferdinand Tönnies to categorize social ties or networks. Today it is defined by many as urban versus rural communities.

4. John McPhee, *Assembling California,* 6–11.

5. McPhee, *Assembling California,* 12–39; Brian J. Sommers, *The Geography of Wine: How Landscapes, Cultures, Terroir, and the Weather Make a Good Drop.*

6. Matt Kramer, *Making Sense of California Wine.* Kramer created the term *somewhereness.*

7. Pinney, *From the Beginnings to Prohibition,* 233–38.

8. Pinney, *From the Beginnings to Prohibition*, 373–81.

9. Paul W. Gates, *California Ranchos and Farms, 1846–1862: Including the Letters of John Quincy Adams Warren of 1861, Being Largely Devoted to Livestock, Wheat Farming, Fruit Raising, and the Wine Industry*, 64; Charles L. Sullivan, *A Companion to California Wine: An Encyclopedia of Wine and Winemaking from the Mission Period to the Present*, 298–99, 168, 171–73. A good reference for the history of the Los Angeles industry can be found in Thomas Pinney, *The City of Vines: A History of Wine in Los Angeles*.

10. Gates, *California Ranchos and Farms*, 64; Vincent P. Carosso, *The California Wine Industry: A Study of the Formative Years*, 7–8; Irving McGee, "Jean Paul Vignes, California's First Professional Winegrower," 176–81.

11. Gates, *California Ranchos and Farms*, 66.

12. Lukacs, *American Vintage*, 63, 69.

13. Adams, *The Wines of America*, 20; Carosso, *California Wine Industry*, 86–102.

14. Eric E. Lampard, *The Rise of the Dairy Industry in Wisconsin, 1820–1920*.

15. Lukacs, *American Vintage*, 57.

16. Maynard A. Amerine, "The Napa Valley Grape and Wine Industry," 289–91.

17. Lukacs, *American Vintage*, 169.

18. Victor W. Geraci, "The El Cajon, California, Raisin Industry: An Exercise in Gilded Age Capitalism," 221–33.

19. For a complete story of the role of Italians in the California wine industry, refer to Simone Cinotto, *Soft Soil, Black Grapes: The Birth of Italian Winemaking in California*.

20. Edmund A. Rossi, "Italian Swiss Colony and the Wine Industry," oral history, 1–9; Deanna Paoli Gumina, "Andrea Sbarboro, Founder of the Italian Swiss Colony Wine Company: Reminiscences of an Italian American Pioneer," 95–160.

21. Doris Muscatine, Maynard A. Amerine, and Bob Thompson, eds., *The University of California/Sotheby Book of California Wine*, 383, 414, 419.

22. Pinney, *From the Beginnings to Prohibition*, 327–31.

23. Discussions of Gilded Age agriculture can be found in Lampard, *Rise of the Dairy Industry in Wisconsin*; and in Allan G. Bogue, *From Prairie to Corn Belt: Farming on the Illinois and Iowa Prairies in the Nineteenth Century*.

24. Geraci, *Wine by Design*.

25. Lukacs, *American Vintage*, 47–57. A discussion of the CWA can be found in Ernest P. Peninou and Gail G. Unzelman, *The California Wine Association and Its Member Wineries, 1894–1920*.

26. Lukacs, *American Vintage*, 57–60 (quote on 59).

27. Muscatine, Amerine, and Thompson, *University of California/Sotheby Book of California Wine*, 383, 414, 419; Pinney, *From the Beginnings to Prohibition*, 374.

28. Erica Hannickel, *Empire of Vines: Wine Culture in America*, 5–13.

CHAPTER 3. LOSS AND REBIRTH OF AN AMERICAN WINE CULTURE

1. A good general discussion of the causes and effects of Prohibition can be found in William J. Rorabaugh, *Prohibition: A Concise History*.

2. Cornelius S. Ough. *Winemaking Basics,* 291–99.

3. Rossi, "Italian Swiss Colony and the Wine Industry," oral history.

4. Norbert C. Mirassou and Edmund A. Mirassou, "The Evolution of a Santa Clara Valley Winery," oral history; Robert Di Giorgio and Joseph A. Di Giorgio, "The Di Giorgios: From Fruit Merchants to Corporate Innovators," oral history.

5. Muscatine, Amerine, and Thompson, *University of California Sotheby Book of California Wine,* 198, 51.

6. Muscatine, Amerine, and Thompson, *University of California Sotheby Book of California Wine,* 62.

7. Peele, "Conflict Between Public Health Goals and the Temperance Mentality," 805–10.

8. Lukacs, *American Vintage,* 103.

9. Seymour Berkson to T. V. Ranck, Rome, December 11, 1932, Than Van Ranck Collection, Sterling Memorial Library.

10. Lukacs, *American Vintage,* 77.

11. Lukacs, *American Vintage,* 106.

12. Lukacs, *American Vintage,* 107.

13. Lukacs, *American Vintage,* 87.

CHAPTER 4. REBUILDING THE WINE INDUSTRY

1. De Luca labeled vintners like Jon Moramarco Sr., Jean Wente, Robert D. Rossi Jr., Al Cribari, Cecil Aguirre, Abraham Buchman, Louis Fappiano Sr., Harold Paul Olmo, Eugene Pio Seghesio, Gene Cuneo, Jim Concannon, and Gene Guglielmo the "Phoenix Generation" during a luncheon speech celebrating the seventieth anniversary of the passage of repeal. Carol Emert, "Vintners Gather in S. F. to Recall the Worst of Times," *San Francisco Chronicle,* February 27, 2003.

2. Louis A. Petri, "The Petri Family in the Wine Industry," oral history, 8.

3. Mirassou and Mirassou, "Evolution of a Santa Clara Valley Winery," oral history, 3–6.

4. Pinney, *From the Beginnings to Prohibition,* 327–31.

5. DiGiorgio and DiGiorgio, "DiGiorgios: From Fruit Merchants to Corporate Innovators," 128–59; Horace O. Lanza, "California Grape Products and Other Wine Enterprises," oral history; Antonio Perelli-Minetti, "A Life in Wine Making," oral history.

6. Ernest Gallo and Julio Gallo, with Bruce B. Henderson, *Ernest and Julio: Our Story,* 147–48.

7. "Resolution Seeks Eastern Trade in California Wines," *The Fresno Bee,* March 21, 1937; "Vintners Seek High Court Ban on Trade Bars," *The Fresno Bee,* January 15, 1938.

8. "Vintners Will Plan Industry Expansion Move," *The Fresno Bee,* October 19, 1934; "Vintners Seek U.S. Funds for Control Program," *The Fresno Bee,* May 13, 1938; "California Wine Men Plan Exhibit at Bay City Fair," *The Fresno Bee,* August 13, 1937.

9. "Move to Ban Low Quality Wine Backed," *Santa Rosa (CA) Press Democrat,* March 31, 1935.

10. "Prospects for Vineyardists to Be Topic at San Jose Meet," *Santa Rosa (CA) Press Democrat,* July 11, 1935.

11. "California Wine Producers Win Long Fight for Standards," *Santa Rosa (CA) Press Democrat,* September 11, 1938.

12. "Wine Men Launch $2,000,000 Drive," *San Francisco Examiner,* June 6, 1939.

13. Pinney, *From the Beginnings to Prohibition,* 114–16.

14. "Trend of Wine Consumption Turning Back to Dry Wines," *Santa Rosa (CA) Press Democrat,* June 27, 1937.

15. "Vintners to Hear Expert on Industry," *Santa Rosa (CA) Press Democrat,* April 24, 1938.

16. Leon Adams, "California Wine Industry Affairs: Recollections and Opinions," oral history, 19–20.

17. Kate Heyhoe and Stanley Hock, *Harvesting the Dream: The Rags-to-Riches Tale of the Sutter Home Winery,* 60.

18. Peyser Jefferson, "The Law and the California Wine Industry," oral history, 6–7.

19. Adams, "California Wine Industry Affairs," oral history.

20. Rossi, "Italian Swiss Colony and the Wine Industry," oral history.

21. Philo Biane, "Wine Making in Southern California and Recollections of Fruit Industries, Inc.," oral history.

22. James Suckling and Jo Cooke, "On the Cutting Edge in an Ancient Land," 45.

23. Suckling and Cooke, "On the Cutting Edge in an Ancient Land," 45–50.

24. Lukacs, *American Vintage,* 111.

25. Charles L. Sullivan, *Napa Wine: A History from Mission Days to Present,* 234–305.

26. Alan E. Fusonie, "John H. Davis: His Contributions to Agricultural Education and Productivity."

27. Pinney, *From the Beginnings to Prohibition,* 347.

28. Richard Bunce, "From California Grapes to California Wine: The Transformation of an Industry, 1963–1979," 57.

29. "C.W.A. Quits Wine Group," *Santa Rosa (CA) Press Democrat,* August 13, 1953.

30. *U.S. News and World Report: The Wine Marketing Handbook, 1972,* 14–15.

31. Adams, "California Wine Industry Affairs," 2.

32. "Leon Adams, Manager of Wine Institute, Quits," *Santa Rosa (CA) Press Democrat,* September 2, 1954.

33. "California FAA Office to Help Wine Men Seen," *Santa Rosa (CA) Press Democrat,* November 26, 1935.

34. Nathan Chroman, "Toasting 1975, a Fine Wine Year," *Los Angeles Times,* August 22, 1976.

35. Donald K. White, "Munro Says: 'Wine Industry Over-Regulated,'" *San Francisco Examiner,* May 15, 1957.

36. "Grape Men Score Tax on Wines," *Santa Rosa (CA) Press Democrat,* March 8, 1951.

37. "Lobbyists' Social Spending Soars," *San Francisco Chronicle,* April 24, 1961.

38. "Political Contributions Questioned: Wine Institute Audit Sought," *San Francisco Examiner,* July 10, 1975.

39. Thomas Pinney, *A History of Wine in America: From Prohibition to the Present,* 357–58.

40. John De Luca, interview with the author, May 2019.

CHAPTER 5. A PERFECT MAN FOR THE JOB

1. John De Luca, in his April 11, 2019, interview with the author, recalled seeing the documents discussing the export of people policy while he was in Rome researching his doctoral dissertation.

2. De Luca, "President and CEO of the Wine Institute," oral history, 305.

3. Simone Cinotto, *The Italian American Table: Food, Family, and Community in New York City,* 2–7.

4. Justin A. Nystrom, *Creole Italian: Sicilian Immigrants and the Shaping of New Orleans Food Culture.*

5. Nystrom, *Creole Italian.*

6. De Luca, interview with the author, April 2019.

7. "Nixon and Khrushchev Argue in Public as U.S. Exhibit Opens; Accuse Each Other of Threats," *The New York Times,* July 24, 1959; Charles Mohr, "Remembrances of the Great 'Kitchen Debate,'" *The New York Times,* July 25, 1984; William Safire, "The Cold War's Hot Kitchen," *The New York Times,* July 23, 2009.

8. De Luca, interview with the author, May 2019.

9. John A. De Luca letter delivered through diplomatic pouch and addressed to "Mr. & Mrs. Peter De Luca & Family," dated September 4, 1961. Letter in possession of the De Luca family.

10. De Luca letter delivered through diplomatic pouch.

11. De Luca, interview with the author, July 2019.

12. De Luca, interview with the author, July 2019.

13. De Luca, interview with the author, July 2019.

14. Russ Cone, "Alioto Cleared of a Conflict over PFEL Tie," *San Francisco Examiner,* September 11, 1974.

15. De Luca, interview with the author, July 2019.

16. De Luca, interview with the author, July 2019.

17. De Luca, interview with the author, July 2019.

18. De Luca, interview with the author, July 2019.

19. De Luca, interview with the author, July 2019.

20. De Luca, interview with the author, July 2019.

21. Carl Cannon, "Wine Institute Chief Faces Headaches," *Los Angeles Times,* August 29, 1975; Nathan Chroman, "Wine Institute Loses Two Members," *Los Angeles Times,* July 24, 1975, 87.

22. De Luca, interview with the author, May 2019.

CHAPTER 6. STABILIZING THE WINE INSTITUTE

1. De Luca, interview with the author, May 2019.

2. De Luca, interview with the author, May 2019.

3. De Luca, interview with the author, May 2019.

4. De Luca, interview with the author, May 2019.

5. De Luca, interview with the author, May 2019.

6. De Luca, interview with the author, May 2019.

7. Adams, *The Wines of America.*

8. De Luca, interview with the author, May 2019.

9. Andrea Sbarboro, *Temperance vs. Prohibition: Important Letters and Data from Our American Consuls, the Clergy and Other Eminent Men,* 13.

10. Bruce Yandle, "Bootleggers and Baptists: The Education of a Regulatory Economist," vii.

11. Bruce Yandle and Adam Smith, *Bootleggers and Baptists: How Economic Forces and Moral Persuasion Interact to Shape Regulatory Practice.*

12. De Luca, "President and CEO of the Wine Institute," oral history, 41–42.

13. Jack Daniels, "Jungle of Rules on Wine Labeling," *The Salt Lake Tribune,* October 16, 1977; Harvey Steinman, "Wine Labels: Can Government Require Too Much Information?," *San Francisco Examiner,* April 18, 1979.

14. "Wine Labeling Plan Opposed at Hearing," *Santa Rosa (CA) Press Democrat,* February 9, 1977; "Government Proposes," *Los Angeles Times,* February 9, 1977: "Planned Government Label Standards for Wine Opposed," *Van Nuys (CA) Valley News,* February 9, 1977, 35; Ruth Ellen Church, "Wine Institute Fights Label Change," *Orlando Sentinel,* November 17, 1977.

15. "New Wine Regulations Will Increase Prices," *Star-Gazette,* November 14, 1977.

16. "Wine Labeling Plan Opposed at Hearing," *Santa Rosa Press (CA) Democrat,* February 9, 1977.

17. "Overly True Wine Labels Could Mislead, Panel Told," *Los Angeles Times,* November 2, 1977.

18. John De Luca, "Remarks by John De Luca; ATF Hearing in San Francisco, February 9, 1977," in De Luca, "President and CEO of the Wine Institute," oral history.

19. Carl Cannon, "Wine Labels: Better Reading but at Higher Prices," *Los Angeles Times,* November 13, 1977.

20. Nathan Chroman, "The Thorny Issue of Wine Labeling," *Los Angeles Times,* October 6, 1977.

21. "Overly True Wine Labels Could Mislead, Panel Told."

22. Carl Cannon, "U.S. 'Seal of Approval' Plan for Domestic Wine Shelved," *Los Angeles Times,* June 15, 1977.

23. De Luca, "President and CEO of the Wine Institute," oral history, 88–89.

24. John A. De Luca, "Code of Advertising," in De Luca, "President and CEO of the Wine Institute," oral history.

25. Carl Cannon, "California Wine Industry Sets Voluntary Ad Code," *Austin (TX) American-Statesman,* April 9, 1978: "Wine Industry Passes Advertising Standards," *Jackson (MS) Clarion-Ledger,* April 10, 1978; Carl Cannon, "Winemakers Adopt Advertising Code to Discourage Abuse of Their Product," *Louisville Courier-Journal,* April 11, 1978.

26. De Luca, "Code of Advertising."

27. Frank J. Prial, "Wine Institute Establishes Standards," *Tampa Tribune,* May 11, 1978.

28. Prial, "Wine Institute Establishes Standards"; Frank J. Prial, "Wine Industry Adopts Strict Guidelines for TV Advertising," *Miami News,* April 28, 1978.

29. De Luca, "President and CEO of the Wine Institute," oral history, 93.

30. De Luca, "President and CEO of the Wine Institute," oral history, 68.

31. John A. De Luca, "Declaration of Principles," in De Luca, "President and CEO of the Wine Institute," oral history.

32. Eric Brazil, "California Seeks New Wine Image," *Lansing (MI) State Journal,* October 18, 1978; Eric Brazil, "California Wineries Cultivate New Image," Richmond, Indiana, *Palladium-Item,* October 19, 1978.

CHAPTER 7. NEOPROHIBITION

1. G. H. Miller and N. Agnew, "The Ledermann Model of Alcohol Consumption: Description, Implications and Assessment"; Eric Single and Victor E. Leino, "The Levels, Patterns, and Consequences of Drinking," in *Drinking Patterns and Their Consequences,* ed. Marcus Grant and Jorge Litvak.

2. John A. De Luca, "The New Prohibitionists: What They Say and How They Affect Legislation and Government Policy," in De Luca, "President and CEO of the Wine Institute," oral history.

3. John A. De Luca, "Wine and Government," Wine Institute Wine Media Day release, June 23, 1978, in De Luca, "President and CEO of the Wine Institute," oral history.

4. De Luca, "The New Prohibitionists."

5. De Luca, "The New Prohibitionists."

6. De Luca, "The New Prohibitionists."

7. John A. De Luca, remarks to Working Group on Prevention, Education, Information, and Training of the Interagency Committee on Federal Activities for Alcohol Abuse and Alcoholism, in De Luca, "President and CEO of the Wine Institute," oral history.

8. John A. De Luca, "The Progress of Wine in America," speech to the Society of Medical Friends of Wine, March 2, 1977, in De Luca, "President and CEO of the Wine Institute," oral history.

9. Nathan Chroman, "Role of California Wine Institute: Two-Day Symposium Scheduled with Medical Groups," *Los Angeles Times,* October 29, 1981.

10. De Luca, "President and CEO of the Wine Institute," 97.

11. De Luca, "President and CEO of the Wine Institute," 97.

12. John De Luca, "Fine Wine and True Grit," document held by the De Luca family.

13. A discussion of the growth of California cuisine and California wine can be found in Victor W. Geraci, *Making Slow Food Fast in California Cuisine.*

14. William J. Eaton, "Liquor Bottlers Ordered to List All Ingredients," *Los Angeles Times,* June 11, 1980.

15. Andrew J. Glass, "Washington Out of Step with Neo-prohibitionists," *Danville (KY) Advocate-Messenger,* January 13, 1985.

16. Peter Navarro, "What Have They Done to the Wine?," *The Washington Post,* January 4, 1982.

17. John De Luca, "Setting the Record Straight on Wine," in De Luca, "President and CEO of the Wine Institute," oral history.

18. De Luca, "Setting the Record Straight on Wine."

19. Howard G. Goldberg, "U.S. Tests for Tainted Wines Continue," *The New York Times,* October 16, 1982.

20. De Luca, "Setting the Record Straight on Wine."

21. Robert Lewis Thompson, "Wining, Divining the French: Two-Hour Tasting at U.S. Embassy Proves Fruitful," *Los Angeles Times,* July 1, 1982; Robert Lewis Thompson, "*Sacre bleu!* California Wineries Test the Waters at a Paris Tasting," *Chicago Tribune,* July 18, 1982.

22. "Wine Institute Sponsors Tasting: 120 Wineries to Pour," *Santa Rosa (CA) Press Democrat,* June 8, 1983.

23. Carl Cannon, "Seagram Agrees to Buy Coke's Wine Properties," *Los Angeles Times,* September 27, 1983.

24. Robert L. Thompson, "Dim Outlook for Wine Growers: California Industry Confronted by Many Problems," *Los Angeles Times,* May 26, 1983.

25. Katherine Brewster, "'Equivalency' Ads Resulting in Wine Activism," *Akron Beacon Journal,* July 16, 1985.

26. Frank J. Prial, "Brawl Between Wine Industry, Distillers Widens Ad Rift," *Longview (TX) News-Journal,* June 19, 1985.

27. Steve Massey, "Distillers Critique Excise-Tax Increase as Latest Attack on Troubled Industry," *Louisville Courier-Journal,* September 29, 1985.

28. Frank J. Prial, "Seagram's, Wine Institute Uncork Sour Vintage," *Minneapolis Star Tribune,* August 28, 1986.

29. Pinney, *From Prohibition to the Present,* 351.

30. Charles Bullard, "Institute Advocates Private Wine Sales," *The Des Moines Register,* October 11, 1983.

31. Stephen Chapman, "Safety Nazis Want to Ban TV Beer Ads," *Orlando Sentinel,* January 27, 1985.

32. John A. De Luca, remarks to Senate Subcommittee on Alcoholism and Drug Abuse, February 7, 1985, in De Luca, "President and CEO of the Wine Institute," an oral history.

33. Bill Keller, "Defending Against Neoprohibitionists," *San Francisco Examiner,* December 6, 1982.

34. Thomas Gephardt, "Here Comes Neo-prohibition," *Cincinnati Enquirer,* August 4, 1985; Andrew Glass, "Industry Trying to Buck Trend: Neoprohibitionists Getting Results," *Dayton (OH) Daily News,* January 14, 1985.

35. De Luca, "President and CEO of the Wine Institute," oral history, 203–4.

CHAPTER 8. ALLA VOSTRA SALUTE—TO YOUR HEALTH

1. "Feds May Enlarge Warning Labels on Wine Bottles," *Wine Spectator,* July 31, 2001.

2. Laurie Woolever, "U.S. Government Mulls Wine Nutrition Labels," *Wine Spectator,* October 15, 2007.

3. Dana Nigro, "The Wine Wars: Tide Turns in Direct Shipping," *Wine Spectator,* October 15, 2002.

4. Geoffrey Mohan, "Overwarned, Underinformed," *Los Angeles Times,* July 26, 2020.

5. Elliot Diringer, "Prop. 65 Science Panel Experts Link Booze to Cancer," *San Francisco Chronicle,* April 23, 1988.

6. De Luca, "President and CEO of the Wine Institute," oral history, 63.

7. De Luca, "President and CEO of the Wine Institute," oral history, 63.

8. De Luca, "President and CEO of the Wine Institute," oral history, 99.

9. Nancy Olson, *With a Lot of Help from Our Friends: The Politics of Alcoholism,* 407–9.

10. De Luca, "President and CEO of the Wine Institute," oral history, 399.

11. M. B. Christie, "Winemakers Circle Their Wagons," *San Francisco Chronicle,* June 25, 1989.

12. De Luca, "President and CEO of the Wine Institute," oral history, 393.

13. John De Luca, "Health and Safety Index," 1986 address to a Symposium on Wine, Health, and Society, in De Luca, "President and CEO of the Wine Institute," oral history.

14. Carol McConnell and Malcolm McConnell, *The Mediterranean Diet: Wine, Pasta, Olive Oil, and a Long, Healthy Life;* John De Luca, "President and CEO of the Wine Institute," oral history, 133, 167.

15. "Woman Named to Head 'Neo-prohibitionist' Fight," *Santa Rosa (CA) Press Democrat,* July 25, 1989.

16. De Luca, "President and CEO of the Wine Institute," oral history, 114.

17. Jim Wood, "The Wine Industry Fights Back: Vintners Wage Battle Against the Neoprohibitionists," *San Francisco Examiner,* May 7, 1989.

18. De Luca, "President and CEO of the Wine Institute," oral history, 393.

19. "Putting the Squeeze on the Grape," *Los Angeles Times,* April 28, 1989; Tim Tesconi, "Wine Country on Warpath," *Santa Rosa (CA) Press Democrat,* October 21, 1990.

20. "Tax Panel OKs Tenfold Increase in Tax on Wine: Wine Institute Behind Measure, Which May Add Liquor and Beer," *San Bernardino County Sun,* May 4, 1989.

21. De Luca, "President and CEO of the Wine Institute," oral history, 120.

22. De Luca, "President and CEO of the Wine Institute," oral history, 169.

23. De Luca, "President and CEO of the Wine Institute," oral history, 169, 170.

24. Daniel P. Puzo, "600 Wines Contain Lead, U.S. Tests Find," *Los Angeles*

Times, August 1, 1991; Paul Jacobs, "Winemakers Stung by Suit Agree to Get the Lead Out," *Los Angeles Times,* November 20, 1991.

25. Jacobs, "Winemakers Stung by Suit Agree to Get the Lead Out."

26. Jack Schreibman, "Americans Drink Less Wine," *Lansing (MI) State Journal,* July 6, 1989; Barbara Ensrud, "A Vintage Decade for Wine: Despite Dip in Consumption, Wine Seems Here to Stay," *San Francisco Examiner,* January 3, 1990.

27. David J. Hanson, "Government's Prohibitionists Attack on Alcohol," *Casper (WY) Star-Tribune,* February 16, 1993.

28. Richard Whitmire, "Surgeon General Targets Teen Alcohol Abuse," *Great (MT) Falls Tribune,* December 10, 1991; Jim Scott, "Surgeon General Was No Friend to Moderate, Social Wine Consumers," *Florida Today* (Cocoa, FL), September 14, 1989; Richard Whitmire, "Surgeon General Targets Underage Drinking," *Bridgewater (NJ) Courier-News,* December 31, 1991.

29. Lewis Perdue, "Don't Believe Everything You Hear About Alcohol," *Wilkes-Barre (PA) Citizens' Voice,* November 20, 1992: "Prohibition: California Wine Industry Struggled to Survive," *Wausau (WI) Daily Herald,* November 29, 1993; "California Wine Pioneers Recall Dry Times," *Appleton (WI) Post Crescent,* December 26, 1993; Lewis Perdue, "Here's to the Truth, Mr. Dry: You Lie!," *San Francisco Examiner,* November 9, 1992.

30. E. Scott Reckard, "The New 'Prohibitionism,'" *San Francisco Examiner,* March 27, 1991.

31. John R. Railman, "Between War, Lice and Puritans, Wine Is an Endangered Species," *Bridgewater (NJ) Courier-News,* June 11, 1994: Tom Marquardt and Patrick Darr, "Responsible Drinkers: Beware of 'Neo-prohibitionists,'" *Easton (MD) Star-Democrat,* September 21, 1990.

32. Lawrence M. Fisher, "Wine Lovers and the War on Drink," *Palm Beach (FL) Post,* June 5, 1990; Mark Fisher, "War on Wine Needs Dose of Common Sense," *Dayton (OH) Daily News,* September 5, 1990; "Wine Aficionados Predict a Trend Toward Prohibition," *Detroit Free Press,* May 25, 1990.

33. John Seiler, "Pry My Fingers Off That Crystal Goblet," *Seymour (IN) Tribune,* December 10, 1992.

34. Lawrence M. Fisher, "Wine Lovers Urge Anti-alcohol Moderation," *Pittsburgh Post-Gazette,* May 23, 1990; "Lover of Wine Warns Against Latest Crusade," *Lincoln Journal Star,* May 29, 1990; Lawrence M. Fisher, "Wine Proponents Fear Negative Activity," *Santa Maria (CA) Times,* May 23, 1990.

35. Tim Tesconi, "Pediatrician-Winemaker Blames Politics for New Sober Campaign," *Santa Rosa (CA) Press Democrat,* January 28, 1990.

36. Nikki Meredith, "Are Anti-alcohol Campaigns Going Too Far?," *Vineland (NJ) Daily Journal,* August 11, 1990.

37. Jon Knutson, "Liquor Industry Wants to Show Drinking Has a Place in Society," *The Bismarck Tribune,* February 14, 1989.

38. K. Dun Gifford and Sara Baer-Sinnot, *The Oldways Table: Essays and Recipes from the Culinary Think Tank;* Dan Berger, "Wherever You Turn, There He Is:

The Missionary Hustler, Colorful, Unsinkable, Unavoidable, He's Robert Mondavi, America's Mr. Wine," *Los Angeles Times,* May 30, 1991; Frank J. Prial, "Wine Drinkers in Control, Industry Study Says: Profile Shows Them as Models of Moderation," *Akron Beacon Journal,* February 26, 1992.

39. De Luca, "President and CEO of the Wine Institute," oral history, 445.

40. Michaela Kane Rodeno, *Bubbles to Boardrooms: Serendipitous Stories from Inside the Wine Business.*

41. De Luca, "President and CEO of the Wine Institute," oral history, 144.

42. De Luca, "President and CEO of the Wine Institute," oral history, 145.

43. De Luca, "President and CEO of the Wine Institute," oral history, 151.

CHAPTER 9. BIONUTRITION, PYRAMIDS, AND LABELS

1. John R. Hailman, "Between War, Lice and Puritans, Wine Is an Endangered Species," *Bridgewater (NJ) Courier-News,* November 6, 1994.

2. Lawrence M. Fisher, "Connoisseurs Are Fighting for Their Right to Drink in the Face of Anti-alcohol Campaigns," *Palm Beach (FL) Post,* June 5, 1990.

3. Carolyn Lochhead, "Wine May Get Health Label; Vintners Lobby for Way to Counteract Warnings," *San Francisco Chronicle,* June 25, 1997.

4. John De Luca, Testimony to the Senate Subcommittee on Alcoholism and Drug Abuse, February 7, 1985, in De Luca, "President and CEO of the Wine Institute," oral history.

5. John De Luca, "Health and Safety Index," paper delivered to the Washington, DC, Symposium on Wine, Health, and Society, in De Luca, "President and CEO of the Wine Institute," oral history.

6. Betty Peterkin and Linda Meyers to John De Luca, June 5, 1990, in De Luca, "President and CEO of the Wine Institute," oral history.

7. Malden Nesheim to Clayton Yeutter and Louis Sullivan, May 14, 1990, in De Luca, "President and CEO of the Wine Institute," oral history.

8. John De Luca, testimony to the US Senate Committee on Commerce, Science, and Transportation, April 2, 1992, in De Luca, "President and CEO of the Wine Institute," oral history.

9. Mark Fisher, "War on Wine Needs Dose of Common Sense," *Dayton (OH) Daily News,* September 5, 1990.

10. S. Welsh, C. Davis, and A. Shaw, "A Brief History of Food Guides in the United States" and "Development of the Food Guide Pyramid."

11. De Luca, "President and CEO of the Wine Institute," oral history, 134.

12. De Luca, "President and CEO of the Wine Institute," oral history, 133.

13. Hearing before the Committee on Agriculture in the House of Representatives, October 23, 1991, in De Luca, "President and CEO of the Wine Institute," oral history.

14. John De Luca, "President's Report," Wine Institute, October 15, 1996, document held by the De Luca family.

15. De Luca, "President and CEO of the Wine Institute," oral history, 206.

16. De Luca, "President and CEO of the Wine Institute," oral history, 206.

17. Letter from Enoch Gordis, MD, to John De Luca, February 24, 1995, in De Luca, "President and CEO of the Wine Institute," oral history.

18. Philip R. Lee, MD, to John De Luca, March 23, 1995, in De Luca, "President and CEO of the Wine Institute," oral history.

19. De Luca, "President and CEO of the Wine Institute," oral history, 207.

20. De Luca, "President and CEO of the Wine Institute," oral history, 204.

21. De Luca, "President and CEO of the Wine Institute," oral history, 126.

22. De Luca, "President and CEO of the Wine Institute," oral history, 195.

23. De Luca, "President and CEO of the Wine Institute," oral history, 248; "In Vino Veritas," *St. Louis Post-Dispatch,* July 28, 1997; Carl Nolte, "Wine Institute in New Guidelines a Drink with Meals Now Called OK," *San Francisco Chronicle,* January 4, 1996.

24. Hilary Abramson, "How Federal Dietary Guidelines Became a Marketing Promotion for Wine," *San Francisco Examiner,* July 5, 1996.

25. Hilary Abramson, "A Song and Dance About Wine: The Government's New Nutritional Guidelines: Verge on Recommending Alcohol Consumption, Shocking the Committee That Wrote Them," *Palm Beach (FL) Post,* July 9, 1996.

26. Carl Hulse, "Wine's Advocates: Vintners' Lobby Comes of Age in Washington," *Santa Rosa (CA) Press Democrat,* December 1, 1996.

27. David Armstrong, "Wine Institute Becomes Industry Power," *Rochester (NY) Democrat and Chronicle,* March 29, 1999.

28. Jim Scott, "Swords Cross Over Labels Sharing Health Benefits of Wine," *Florida Today,* March 11, 1999.

29. "Wine Pioneers Remember the 'Dry Times,'" *The Newark (OH) Advocate,* January 29, 1994; "Prohibition: California Wine Industry Struggled to Survive," *Wausau (WI) Daily Herald,* November 29, 1993; "Wine Industry Died on the Vine in 1930s," *Appleton (WI) Post-Crescent,* December 26, 1993.

30. De Luca, "President and CEO of the Wine Institute," oral history, 160.

31. David J. Hanson, "Government's Prohibitionist Attack on Alcohol," *Casper (WY) Star-Tribune,* February 16, 1993, 6.

32. De Luca, "President and CEO of the Wine Institute," oral history, 176.

33. De Luca, "President and CEO of the Wine Institute," oral history, 128–29.

34. Ray Delgado, "Wineries Say They Have Secret Weapon in Label War: Research," *San Francisco Examiner,* December 14, 1999.

35. Stuart Auerbach, "Wine Industry Tries to Promote Health," *Bridgewater (NJ) Courier-News,* May 12, 1996; Ted Appel, "Vintners Urge New Wine Label: Request to List Health Benefits, Too," *Santa Rosa (CA) Press Democrat,* May 7, 1996.

36. Martha Groves, "Sour News for Wine: New Health Report Is No Seal of Approval," *Los Angeles Times,* March 5, 1998.

37. "To Your Health! S.F.'s Wine Institute Qualifies for the Emperor Norton Prize by Seeking Labels That Depict Dr. Bacchus as a Health Provider," *San Francisco Examiner,* May 13, 1996. The Emperor Norton Awards are satirical San Francisco Bay Area specific awards given for "extraordinary invention and

creativity unhindered by the constraints of paltry reason." The award is named after Joshua Norton I, who in 1859, proclaimed himself to be the emperor of the United States of America and protector of Mexico.

38. Richard Mendelson, *From Demon to Darling: A Legal History of Wine in America,* 168–71.

39. Mendelson, *From Demon to Darling,* 170; "Health: U.S. to Propose Changes in Wine Labeling," *Los Angeles Times,* February 5, 1999.

40. Michael Boyle, "Strom Thurmond Re-opens Wine-Label Issue," *Santa Maria (CA) Times,* May 27, 1999.

41. John De Luca, "Federal Government Approves Dietary Guidelines Label Statement for Wine," Wine Institute memo and news release, in De Luca, "President and CEO of the Wine Institute," oral history.

42. Ken Marcus, "Strom Thurmond's Crusade Against Wine," *Wine Spectator,* July 31, 1999, 38–44.

43. "Thurmond Calls for Wine Institute Investigation," *Wine Business Monthly* (April 1999): 8; Marcus, "Strom Thurmond's Crusade Against Wine."

44. George Raine, "A Pressing Problem; Strom Thurmond Targets Vintners," *San Francisco Examiner,* August 9, 1999; Michael Doyle, "ATF Puts Cork in Labels Touting Wine's Health Benefits," *Santa Maria (CA) Times,* December 15, 1999.

45. Strom Thurmond to Roger Viadero, February 26, 1999, in De Luca, "President and CEO of the Wine Institute," oral history.

46. De Luca, "President and CEO of the Wine Institute," oral history, 269.

47. Gerald D. Boyd, "New Wine Label Causes Uproar," *San Francisco Chronicle,* March 17, 1999.

48. De Luca, "President and CEO of the Wine Institute," oral history, 256.

CHAPTER 10. DIRECT SHIPPING

1. De Luca, "President and CEO of the Wine Institute," oral history, 276.

2. Frank J. Prial, "Wine Talk; Bottles by Mail, from Brooks Brothers," *New York Times,* October 14, 1998; Michelle Slatalla, "Online Shopper; Reds and Whites, and Shipping Blues," *The New York Times,* March 30, 2000.

3. R. W. Apple Jr., "Zinfandel by Mail? Well, Yes and No; Strict Laws May Get Stricter," *New York Times,* May 19, 1999.

4. De Luca, "President and CEO of the Wine Institute," oral history, 292.

5. De Luca, "President and CEO of the Wine Institute," oral history, 294.

6. Article I, Section 8, Clause 3, of the United States Constitution, the Commerce Clause, states that the United States Congress shall have power "To regulate Commerce with foreign Nations, and among the several States, and with the Indian Tribes." The Interstate Commerce Clause has been laden with political controversy because it helps define relationships between states and the power of the executive and legislative branches to intervene.

7. Orrin Hatch, "Statement of Senator Orin Hatch, Senate Judiciary Committee, Hearing on Interstate Alcohol Sales and the 21st Amendment," March 9, 1999, in De Luca, "President and CEO of the Wine Institute," oral history.

8. De Luca, "President and CEO of the Wine Institute," oral history, 294.

9. De Luca, "President and CEO of the Wine Institute," oral history, 280–81.

10. Cyril Penn, "Direct Shipping Issue Gains Increasing Visibility," 43–44; "Key Senator Supports States in Crackdown on Home Delivery of Wine," *Wine Spectator,* May 15, 1999.

11. "Wineries," *The Ukiah (CA) Daily Journal,* March 1, 1999.

12. Clyde Weiss, "Sen. Thurmond Asks for Investigation of Wine Institute," *The Ukiah (CA) Daily Journal,* March 2, 1999.

13. "Wine," *Los Angeles Times,* June 14, 1999.

14. "Feinstein and Wilson Get Involved in Direct Shipping Debate," *Wine Business Monthly* (November 1998): 9.

15. Dana Nigro, "Home Delivery Crackdown Approved by Wide Margin," *Wine Spectator,* September 15, 1999.

16. "Key Senator Supports States in Crackdown on Home Delivery of Wine."

17. John De Luca and Wine Institute, "Wine Industry Code for Direct Shipping," San Francisco, January 12, 1999, in De Luca, "President and CEO of the Wine Institute," oral history.

18. Nigro, "Home Delivery Crackdown Approved by Wide Margin."

19. Nick Anderson, "Battle Brews over Online Sales of Alcohol," *Los Angeles Times,* June 14, 1999.

20. De Luca, "President and CEO of the Wine Institute," oral history, 291–97.

21. Dana Nigro, "Home Delivery Battle," *Wine Spectator,* August 31, 1999.

22. Ted Appel, "K-J Suit Takes on Illinois Law," *Santa Rosa (CA) Press Democrat,* June 10, 1999.

23. De Luca, "President and CEO of the Wine Institute," oral history, 278.

24. De Luca, "President and CEO of the Wine Institute," oral history, 294.

25. "Wine Industry Code for Direct Shipping," in De Luca, "President and CEO of the Wine Institute," oral history.

26. Tim Fish and Dana Nigro, "Wine-Shipping Battle Heats Up in Federal Courts," *Wine Spectator,* June 15, 2002.

27. Dana Nigro, "Federal Judge Overturns Texas Shipping Laws," *Wine Spectator,* September 15, 2002.

28. "California Debates Restricting Wine Imports," *Wine Spectator,* June 30, 2002; "California Bill to Restrict Wine Imports Stalls," *Wine Spectator,* July 31, 2002.

29. Dana Nigro, "Courts Favor Consumers in New York and Florida Wine-Shipping Lawsuits," *Wine Spectator,* January 31, 2003; Dana Nigro, "Federal Appeals Court Nixes Michigan's Wine-Shipping Ban," *Wine Spectator,* November 30, 2003.

30. Dana Nigro, "From Whitewater to Wine: Kenneth Starr Joins Shipping Fight," *Wine Spectator,* May 15, 2003.

31. Dana Nigro, "High Court to Hear Wine Shipping Case," *Wine Spectator,* July 31, 2004.

32. "Uncork Free Trade," *San Francisco Chronicle,* December 13, 2004; Michael Doyle, "Wineries Unite on Direct Shipping," *The Sacramento Bee,* December 11,

2004; Richard Wilner, "New York Wine Lovers May Save Big in 2005," *New York Post,* December 9, 2004; Alan Goldfarb, "The Wine Industry's Case for Direct Shipping," *St. Helena (CA) Star,* December 9, 2004; Linda Greenhouse, "Justices Pick Apart Ban on Wine Sales from State to State," *The New York Times,* December 8, 2004; Robert S. Greenberger and Mark H. Anderson, "Supreme Court Seems Open to Sales by Out-of-State Wineries," *The Wall Street Journal,* December 8, 2004; "Vintners Optimistic After Hearing Questions by Supreme Court Justices Seemed to Favor Winemaker's Case," *Santa Rosa (CA) Press Democrat,* December 8, 2004.

33. Stephen Evans, "Prohibition Still Hurts America's Wine Makers," BBC News, December 8, 2004.

34. Nick Fauchald and Dana Nigro, "Online Wine Sellers Caught in a Sting," *Wine Spectator,* August 31, 2004.

35. A legal case study of the Supreme Court decision can be found in Carol Robertson, "Case 9: You Can't Take It with You, or Have It Shipped Either—Direct Shipment: The Supreme Court Weighs In," in *The Little Red Book of Wine Law: A Case of Legal Issues,* 104–19.

36. Dana Nigro, "U.S. Supreme Court Overturns Wine-Shipping Bans," *Wine Spectator,* May 16, 2005; "Justices Lift Ban on Wine Shipments," *Santa Barbara News-Press,* May 17, 2005; Marvin R. Shanken, "A Supreme Decision," *Wine Spectator,* July 31, 2005.

37. Nigro, "U.S. Supreme Court Overturns Wine-Shipping Bans."

CHAPTER 11. POLITICS OF WINE

1. De Luca, "President and CEO of the Wine Institute," oral history, 218.

2. De Luca, "President and CEO of the Wine Institute," oral history, 218.

3. De Luca, "President and CEO of the Wine Institute," oral history, 219.

4. De Luca, "President and CEO of the Wine Institute," oral history, 346–47.

5. De Luca, "President and CEO of the Wine Institute," oral history, 347.

6. De Luca, "President and CEO of the Wine Institute," oral history, 347.

7. De Luca, "President and CEO of the Wine Institute," oral history, 396.

8. Nick Anderson, "The Washington Connection: State Delegation Uncorks Wine Politics," *Los Angeles Times,* April 27, 1999.

9. Anderson, "The Washington Connection."

10. De Luca, "President and CEO of the Wine Institute," oral history, 150.

11. De Luca, "President and CEO of the Wine Institute," oral history, 288.

12. https://winecaucus-mikethompson.house.gov/about.

13. https://winecaucus-mikethompson.house.gov/issues.

CHAPTER 12. ESTABLISHING THE PATH FOR THE WINE INSTITUTE'S FUTURE

1. De Luca, "President and CEO of the Wine Institute," oral history, 335.

2. Jefferson C. M. Coombs, president the Alumni Associations of the University of California, to John De Luca, PhD, March 14, 2007, document held by the De Luca family.

3. Jennifer O'Brien, "Gallo Research Center to Lead $15 Million U.S. Army Funded National Research Program," UCSF press release, July 11, 2012, document held by the De Luca family.

4. De Luca, "President and CEO of the Wine Institute," oral history, 336.

5. De Luca, "President and CEO of the Wine Institute," oral history, 336–38.

6. De Luca, "President and CEO of the Wine Institute," oral history, 400.

7. De Luca, "President and CEO of the Wine Institute," oral history.

EPILOGUE. WHO WILL CARRY THE BANNER NEXT?

1. De Luca, "President and CEO of the Wine Institute," oral history, 400.

2. De Luca, "President and CEO of the Wine Institute," oral history, 403.

3. Lewis Perdue, "How the Wine Institute Fumbled the Moderate Consumption Fight and Aided Neoprohibition," *Wine Business Monthly,* June 20, 2019.

4. Barry Cameron, "Can a Christian Drink Alcohol?" *Church Leaders,* 26 July 2018.

5. Paul D. Shinkman, "Study: No Amount of Alcohol Is Healthy," *U.S. News and World Report,* August 23, 2018; Josh Gabbatiss, "No Safe Level of Alcohol Consumption, Major Study Concludes," *The Independent,* August 23, 2018; Nicholas Bakalar, "How Much Alcohol Is Safe to Drink? None, Say These Researchers," *The New York Times,* August 27, 2018.

6. Katelyn Newman, "Study: Alcohol Is Fueling an 'Urgent Public Health Crisis' in the U.S.," *U.S. News and World Report,* February 21, 2020; Gaby Galvin, "Too Much Alcohol Has Taken the Deadliest Toll in These States," *U.S. News and World Report,* July 31, 2020; Sumathi Reddy, "Men Urged to Limit Alcohol to One Drink a Day amid New Concerns," *The Wall Street Journal,* August 17, 2020.

Sources Consulted

NEWSPAPERS

Akron Beacon Journal
Appleton (WI) Post Crescent
Austin (TX) American-Statesman
Bismarck Tribune
Bridgewater (NJ) Courier-News
Casper (WY) Star-Tribune
Chicago Tribune
Cincinnati Enquirer
Danville (KY) Advocate-Messenger
Dayton (OH) Daily News
Des Moines Register
Detroit Free Press
Easton (MD) Star Democrat
Escondido (CO) Daily Times Advocate
Florida Today (Cocoa, FL)
Fresno Bee
Great Falls (MT) Tribune
Guardian
Independent
Jackson (MS) Clarion-Ledger
Lansing (MI) State Journal
Lincoln Journal Star
Longview (TX) News-Journal
Los Angeles Herald Examiner
Los Angeles Times
Los Padres Sun (Santa Ynez, CA)
Louisville Courier-Journal
Minneapolis Star Tribune
Modesto (CA) Bee
Napa Valley Register

Newark (OH) Advocate
New York Star-Gazette
New York Times
Orlando Sentinel
Palm Beach (FL) Post
Pittsburgh Post-Gazette
Richmond (IN) Palladium-Item
Rochester (NY) Democrat and Chronicle
Sacramento Bee
Salt Lake Tribune
San Bernardino County Sun
San Francisco Chronicle
San Francisco Examiner
Santa Barbara Independent
Santa Barbara News-Press
Santa Maria (CA) Times
Santa Rosa (CA) Press Democrat
Santa Ynez Valley News (Solvang, CA)
Seymour (IN) Tribune
St. Helena (CA) Star
St. Louis Post-Dispatch
Tampa Tribune
Ukiah (CA) Daily Journal
Van Nuys (CA) Valley News
Vineland (NJ) Daily Journal
Wall Street Journal
Washington Post
Wausau (WI) Daily Herald
Wilkes-Barre (PA) Citizens' Voice

MAGAZINES

Advertising Age
American Heritage
Business Week

California Farmer
California Law Business
California Wine Tasting Monthly

Decanter   *U.S. News and World Report*
Forbes   *Wine Business Monthly*
The New Yorker   *Wine Enthusiast*
Pacific Wine and Vines   *Wine News*
Relax: The Travel Magazine for   *Wines and Vines*
   Practicing Physicians   *Wine Spectator*
Time

ORAL HISTORIES

Adams, Leon. "California Wine Industry Affairs: Recollections and Opinions."
Oral history interview by Ruth Teiser, 1986. Oral History Center, Bancroft
Library, University of California, Berkeley, 1990.

Amerine, Maynard A. "The University of California and the State's Wine Industry." Oral history interview by Ruth Teiser, 1969–71. Oral History Center, Bancroft Library, University of California, Berkeley, 1972.

Biane, Philo. "Wine Making in Southern California and Recollections of Fruit Industries, Inc." Oral history by Ruth Teiser, 1972. Oral History Center, Bancroft Library, University of California, Berkeley.

Ciocca, Arthur A. "Arthur A. Ciocca and the Wine Group, Inc.: Insights into the Wine Industry from a Marketing Perspective." Oral history by Carole Hicke, 1999. Oral History Center, Bancroft Library, University of California, Berkeley.

Critchfield, Burke H., Karl F. Wente, and Andrew G. Frericks. "The California Wine Industry During the Depression." Oral history interview by Ruth Teiser, 1972. Oral History Center, Bancroft Library, University of California, Berkeley, 1972.

Cruess, William V. "A Half Century in Food and Wine Technology." Oral history by Ruth Teiser, 1967. Oral History Center, Bancroft Library, University of California, Berkeley.

De Luca, John. Oral interviews conducted by Victor W. Geraci, May 2019–February 2020. Recordings in the possession of the author.

———. "President and CEO of the Wine Institute, 1975–2003." Oral history by Victor W. Geraci, Ruth Teiser, and Carole Hicke, 1986–2007. Oral History Center, Bancroft Library, University of California, Berkeley.

DiGiorgio, Robert, and Joseph A. DiGiorgio. "The DiGiorgios: From Fruit Merchants to Corporate Innovators." Oral history by Ruth Teiser, 1983. Oral History Center, Bancroft Library, University of California, Berkeley, 1986.

Gallo, Ernest. "The E. & J. Gallo Winery." Oral history by Ruth Teiser, 1969. Oral History Center, Bancroft Library, University of California, Berkeley, 1995.

Gomberg, Louis. "Analytical Perspectives on the California Wine Industry, 1935–1990." Oral history by Ruth Teiser, 1990. Oral History Center, Bancroft Library, University of California, Berkeley, 1990.

Lanza, Horace O. "California Grape Products and Other Wine Enterprises." Oral history interview by Ruth Teiser, 1969. Oral History Center, Bancroft Library, University of California, Berkeley, 1971.

Meyer, Otto E. "California Premium Wines and Brandy." Oral history interview by Ruth Teiser, 1971. Oral History Center, Bancroft Library, University of California, Berkeley, 1973.

Mirassou, Norbert C., and Edmund A. Mirassou. "The Evolution of a Santa Clara Valley Winery." Oral history by Ruth Teiser, 1984. Oral History Center, Bancroft Library, University of California, Berkeley.

Moone, Michael. "Management and Marketing at Beringer Vineyards and Wine World, Inc." Oral history by Lisa Jacobson, 1989. Oral History Center, Bancroft Library, University of California, Berkeley, 1990.

Muscatine, Doris. "Food and Wine Writer." Oral history by Victor W. Geraci, 2004. Oral History Center, Bancroft Library, University of California, Berkeley, 2006.

Olmo, Harold P. "Plant Genetics and New Grape Varieties." Oral history interview by Ruth Teiser, 1972–73. Oral History Center, Bancroft Library, University of California, Berkeley, 1976.

Ough, Cornelius. "Researches of an Enologist University of California, Davis, 1950–1990." Oral history by Ruth Teiser, 1989. Oral History Center, Bancroft Library, University of California, Berkeley, 1990. Taken from the bound oral history held by the Oral History Center.

Perelli-Minetti, Antonio. "A Life in Wine Making." Oral history interview by Ruth Teiser, 1969. Oral History Center, Bancroft Library, University of California, Berkeley, 1975.

Petri, Louis A. "The Petri Family in the Wine Industry." Oral history by Ruth Teiser, 1971. Oral History Center, Bancroft Library, University of California, Berkeley.

Peyser, Jefferson. "The Law and the California Wine Industry." Oral history by Ruth Teiser, 1973. Oral History Center, Bancroft Library, University of California 1974.

Rossi, Edmund A. "Swiss Colony and the Wine Industry." Oral history interview by Ruth Teiser, 1969. Oral History Center, Bancroft Library, University of California, Berkeley, 1971.

Tchelistcheff, André. "Grapes, Wine, and Ecology." Oral history by Ruth Teiser and Catherine Harroun, 1979. Oral History Center, Bancroft Library, University of California, Berkeley, 1983.

OTHER SOURCES

Adams, Leon D. The Commonsense Book of Wine. New York: D. McKay, 1964.
———. The Wines of America. San Francisco: McGraw-Hill, 1990.

Amerine, Maynard A. "An Introduction to the Pre-repeal History of Wine." Agricultural History 63 (April 1969): 259–68.
———. "The Napa Valley Grape and Wine Industry." Agricultural History 49 (Spring 1975): 289–91.

Amerine, Maynard A., and Maynard A. Joslyn. Commercial Production of Table Wines. Berkeley: California Agricultural Experimental Station, College of Agriculture, University of California, 1940.

Amerine, Maynard A., and H. Phaff. *A Bibliography of Publications by the Faculty, Staff, and Students of the University of California, 1876–1980, on Grapes, Wines, and Related Subjects.* Berkeley: University of California Press, 1986.

Amerine, Maynard A., and Vernon L. Singleton. *Wine: An Introduction.* 2nd ed. Berkeley: University of California Press, 1977.

Asher, Gerald. *A Vineyard in My Glass.* Berkeley: University of California Press, 2011.

Baxter, Richard, MD. *Wine and Health: Making Sense of the New Science and What It Means for Wine Lovers.* San Francisco: Board and Bench, 2019.

Bennett, M. K. "Climate and Agriculture in California." *Economic Geography* 15 (April 1939): 158–63.

Best, Michael R. "The Mystery of Vintners." *Agricultural History* 50 (April 1976): 362–76.

Blout, Jesse S. *A Brief Economic History of the California Wine-Growing Industry.* San Francisco: Bureau of Markets, California Department of Agriculture, 1943.

Bogue, Allan G. *From Prairie to Corn Belt: Farming on the Illinois and Iowa Prairies in the Nineteenth Century.* Chicago: Quadrangle Press, 1963.

Braconi, Frank. *The U.S. Wine Market: An Economic Marketing & Financial Investigation.* Merrick, NY: Morton Research Corporation, April 1977.

Breimyer, Harold F. "The Economic Returns of Agricultural Education." *Agricultural History* 60 (Spring 1986): 65–72.

Brenner, Deborah. *Women of the Wine Industry: Inside the World of Women Who Make, Taste, and Enjoy Wine.* Hoboken, NJ: John Wiley & Sons, 2007.

Bunce, Richard. "From California Grapes to California Wine: The Transformation of an Industry, 1963–1979." *Contemporary Drug Problems* (Spring 1981): 55–74.

Burnham, John C. *Bad Habits: Drinking, Smoking, Taking Drugs, Gambling, Sexual Misbehavior, and Swearing in American History.* New York: New York University Press, 1993.

Byles, Stuart Douglass. *Los Angeles Wine: A History from the Mission Era to the Present.* Charleston, SC: American Palate, 2014.

California Agricultural Experiment Station, State Board of Public Health, Bureau of Food and Drug Inspection. *Regulations Establishing Standards of Identity, Quality, Purity and Sanitation, and Governing the Labeling and Advertising of Wine in the State of California.* Sacramento: California State Printing Office, 1942.

California Agricultural Statistics Service. *California Grape Acreage, 1970–2019.* Sacramento: California Agricultural Statistics Service.

Cameron, Barry. "Can a Christian Drink Alcohol?" *ChurchLeaders,* July 26, 2018.

Carosso, Vincent P. *The California Wine Industry: A Study of the Formative Years.* Berkeley: University of California Press, 1951.

Castiglione, G. E. Di Palma. "Italian Immigration into the United States 1901–4." *American Journal of Sociology* 11 (September 1905): 183–206.

Chopas, Mary Elizabeth Basile. *Searching for Subversives: The Story of Italian Internment in Wartime America.* Chapel Hill: University of North Carolina Press, 2017.

Cinotto, Simone. *The Italian American Table: Food, Family, and Community in New York City.* Chicago; University of Illinois Press, 2013.

———. *Soft Soil, Black Grapes: The Birth of Italian Winemaking in California.* New York: New York University Press, 2012.

Cipolla, Carlo. "European Connoisseurs and California Wines." *Agricultural History* 44, no. 1 (1975): 294–310.

Colman, Tyler. *Wine Politics: How Governments, Environmentalists, Mobsters, and Critics Influence the Wines We Drink.* Berkeley: University of California Press, 2008.

Cordasco, Francesco, and Eugene Bucchioni. *The Italians: Social Backgrounds of an American Group.* Clifton, NJ: Augustus M. Kelley, 1974.

Curry, James Harold, III. "Agriculture Under Late Capitalism: The Structure and Operation of the California Wine Industry." PhD diss., Cornell University, 1994.

Dans, Peter, and Suzanne Wasserman. *Life on the Lower East Side, 1937–1950.* Photographs by Rebecca Lepkoff. New York: Princeton Architectural Press, 2006.

de la Peña, Donald Joseph. "Vineyards in a Regional System of Open Space in the San Francisco Bay Area: Methods of Preserving Selected Areas." Master's thesis, University of California, Santa Clara, 1962.

Della Valle, Andrea Laura. "The Taste of Globalization: The Wine Industry of Ontario" Master's thesis, University of Windsor, 1996.

De Luca, John A. "The Italian Road to Socialism." PhD diss., University of California, Los Angeles, 1967.

Dolan, Paul, with Thom Elkjer. *True to Our Roots: Fermenting a Business Revolution.* Princeton, NJ: Bloomberg Press, 2003.

Emelise, Aleandri. *Images of America: Little Italy.* San Francisco: Arcadia, 2002.

Fauntleroy, Phylicia Ann. "An Economic Analysis of the United States Demand for Distilled Spirits, Wine, and Beer Incorporating Taste Changes Through Demographic Factors, 1960–1981 (Beverage Alcohol)." PhD diss., American University, 1984.

Fisher, M. F. K. *The Story of Wine in California.* Berkeley: University of California Press, 1962.

Florence, Jack W. *Legacy of a Village: The Italian Swiss Colony Winery and People of Asti, California.* Phoenix: Raymond Court Press, 1999.

Folwell, Raymond J., and Mark A. Castaldi. "Economies of Size in Wineries and Impacts of Pricing and Product Mix." *Agribusiness: An International Journal* 3 (Fall 1987): 281–92.

Ford, Gene. *To Your Health: The Science of Healthy Drinking.* San Francisco: Wine Appreciation Guild, 2003.

Fuller, Robert C. *Religion and Wine: A Cultural History of Wine Drinking in the United States.* Knoxville: University of Tennessee Press, 1996.

Fusonie, Alan E. "John H. Davis: His Contributions to Agricultural Education and Productivity." *Agricultural History* 60 (Spring 1986): 97–110.

Gallo, Ernest, and Julio Gallo, with Bruce Henderson. *Ernest and Julio Gallo: Our Story*. New York: Random House, 1994.

Gates, Paul W. *California Ranchos and Farms, 1846–1862: Including the Letters of John Quincy Adams Warren of 1861, Being Largely Devoted to Livestock, Wheat Farming, Fruit Raising, and the Wine Industry*. Madison: State Historical Society of Wisconsin, 1967.

Geraci, Victor W. "El Cajon, California, 1900." *Journal of San Diego History* 36 (Fall 1990): 221–33.

———. "The El Cajon, California, Raisin Industry: An Exercise in Gilded Age Capitalism." *Southern California Quarterly* 74 (Winter 1992): 329–54.

———. "The Family Wine-Farm: Vintibusiness Style." *Agricultural History* 74 (Spring 2000): 419–32.

———. "Grape Growing to Vintibusiness: A History of the Santa Barbara, California, Regional Wine Industry, 1965–1995." PhD diss., University of California, Santa Barbara, 1997.

———. *Making Slow Food Fast in California Cuisine*. Cham, Switzerland: Palgrave Macmillan, 2017.

———. "The Rise and Fall of the El Cajon, California, Raisin Industry: 1873–1920." Master's thesis, San Diego State University, 1990.

———. *Salud: The Rise of Santa Barbara's Wine Industry*. Reno: University of Nevada Press, 2004.

———. *Santa Barbara New House: The First Forty Years, 1955–1995*. Santa Barbara, CA: Santa Barbara New House, 1995.

———. "Vintibusiness: The History of the California Wine Industry, 1769 to the Present." In *California History: A Topical Approach,* edited by Gordon Morris Bakken. Wheeling, IL: Harlan Davidson, 2002.

———. "Wine, Women, and Song." In *Encyclopedia of Women in the American West,* Gordon Morris Bakken and Brenda Farrington, eds. Thousand Oaks, CA: Sage, 2003.

———. *Wine by Design: Santa Barbara's Quest for Terroir*. Reno: University of Nevada Press, 2020.

Geraci, Victor W., and Elizabeth S. Demers, eds. *Icons of American Cooking*. Santa Barbara, CA: Greenwood, 2011.

Gifford, K. Dun, and Sara Baer-Sinnot. *The Oldways Table: Essays and Recipes from the Culinary Think Tank*. Berkeley: Ten Speed Press, 2007.

Grant, Marcus, and Jorge Litvak, eds. *Drinking Patterns and Their Consequences*. Philadelphia: Taylor and Francis, 1998.

Gumina, Deanna Paoli. "Andrea Sbarboro, Founder of the Italian Swiss Colony Wine Company: Reminiscences of an Italian American Pioneer." In *Struggle and Success: An Anthology of the Italian Immigrant Experience in California,* edited by Paola A. Sensi-Isolani and Phylis Cancilla Martinelli, eds. 95–106. New York: Center for Immigration Studies, 1993.

Hakluyt, Richard. *The Principal Navigations, Voyages, Traffiques, and Discoveries of the English Nation.* 1589. Reprint, Glasgow, 1903–5.

Hannickel, Erica. *Empire of Vines: Wine Culture in America.* Philadelphia: University of Pennsylvania Press, 2013.

Hawkes, Ellen. *Blood and Wine: The Unauthorized Story of the Gallo Wine Empire.* New York: Simon & Schuster, 1993.

Heien, Dale M. "The Impact of the Alcohol Tax Act of 1990 on California Agriculture." A position paper sponsored by the United Agribusiness League, Irvine, CA, July 1990.

Heyhoe, Kate, and Stanley Hock. *Harvesting the Dream: The Rags-to-Riches Tale of the Sutter Home Winery.* Hoboken, NJ: John Wiley & Sons, 2004.

Hilgard, Eugene W. *Report of the Professor of Agriculture to the President of the University.* Sacramento: State Printing Office, 1879.

———. *University of California—College of Agriculture Report of the Viticultural Work During the Seasons 1887–1893 with Data Regarding the Crush of 1894–95.* Sacramento: Superintendent State Printing, 1896.

Howes, Nina, and Eric Ferrara. *Lower East Side Oral Histories.* Charleston, SC: History Press, 2012.

Hyams, E. *Dionysus: A Social History of the Wine Vine.* 2nd ed. London: Jackson, 1987.

Isrealowitz, Oscar, and Brian Merlis. *Manhattan's Lower East Side in Vintage Photographs.* Brooklyn, NY: Israelowitz, 2009.

Johnson, Hugh. *Vintage: The Story of California Wine.* New York: Simon and Schuster, 1989.

Johnson, Robert. "Nova Britannia." In *Tracts Relating Principally to the Origin, Settlement, and Progress of the Colonies in North America,* Peter Force, ed. 4 vols. Washington, DC, 1836–46.

Jones, Frank. *The Save Your Heart Wine Guide.* New York: St. Martin's Press, 1996.

Jones, Idwal. *Vines in the Sun: A Journey Through the California Vineyards.* New York: Ballantine Books, 1949.

Kladstrup, Don, and Petie Kladstrup, with J. Kim Munholland. *Wine and War: The French, the Nazis, and the Battle for France's Greatest Treasure.* New York: Broadway Books, 2001.

Kolpan, Steven. *A Sense of Place: An Intimate Portrait of the Niebaum-Coppola Winery and the Napa Valley.* New York: Routledge, 1999.

Kramer, Matt. *Making Sense of California Wine.* New York: William Morrow, 1992.

———. *Making Sense of Napa Valley, Sonoma, Central Coast, and Beyond.* Philadelphia: Running Press, 2004.

Lampard, Eric E. *The Rise of the Dairy Industry in Wisconsin, 1820–1920.* Madison: State Historical Society of Wisconsin, 1967.

Lapsley, James T. *Bottled Poetry: Napa Winemaking from Prohibition to the Modern Era.* Berkeley: University of California Press, 1996.

Larsen, John W. *Vineyard Development Financing in California*. San Francisco: Wells Fargo Bank, 1972.

Lewis, Michael, and Richard F. Hamm, eds. *Prohibition's Greatest Myths: The Distilled Truth About America's Anti-alcohol Crusade*. Baton Rouge: Louisiana State University Press, 2020.

Logevall, Fredrik. *Choosing War: The Lost Chance for Peace and the Escalation of War in Vietnam*. Berkeley: University of California Press, 1999.

Lukacs, Paul. *American Vintage: The Rise of American Wine*. New York: Houghton Mifflin, 2000.

———. *Inventing Wine: A New History of One of the World's Most Ancient Pleasures*. New York: W. W. Norton, 2012.

Lynch, Kermit. *Adventures on the Wine Route: A Wine Buyer's Tour of France*. New York: North Point Press, 1988.

Matasar, Ann B. *Women of Wine: The Rise of Women in the Global Wine Industry*. Berkeley: University of California Press, 2006.

McConnell, Carol, and Malcolm McConnell. *The Mediterranean Diet: Wine, Pasta, Olive Oil, and a Long, Healthy Life*. New York: W. W. Norton, 1987.

McCoy, Elin. *The Emperor of Wine: The Rise of Robert Parker, Jr. and the Reign of American Taste*. New York: Ecco, 2005.

McGee, Irving. "The Beginnings of California Winegrowing." *Historical Society of Southern California Quarterly* 29 (March 1947): 59–71.

———. "Early California Winegrowers." *California Magazine of the Pacific* 37 (September 1947): 34–37.

———. "Jean Paul Vignes, California's First Professional Winegrower." *Agricultural History* 22 (1948): 176–81.

McGirr, Lisa. *The War on Alcohol: Prohibition and the Rise of the American State*. New York: W. W. Norton, 2016.

McNutt, Joni G. *In Praise of Wine: An Offering of Hearty Toasts, Quotations, Witticisms, Proverbs, and Poetry Throughout History*. Santa Barbara, CA: Capra Press, 1993.

McPhee, John. *Assembling California*. New York: Farrar, Straus & Giroux, 1993.

Mendelson, Richard. *From Demon to Darling: A Legal History of Wine In America*. Berkeley: University of California Press, 2009.

———. *Wine in America: Law and Policy*. New York: Wolters Kluwer Law & Business, 2011.

Miller, G. H., and N. Agnew. "The Ledermann Model of Alcohol Consumption: Description, Implications and Assessment." *Quarterly Journal of Studies on Alcohol* 35 (September 1974): 877–98.

Mondavi, Robert, with Paul Chutkow. *Harvests of Joy: My Passion for Excellence*. New York: Harcourt Brace, 1998.

Moulton, Kirby, and James Lapsley, eds. *Successful Wine Marketing*. Gaithersburg, MD: Aspen, 2001.

Mouton, Kirby S., ed. *The Economics of Small Wineries: The Proceedings of Two Seminars at University of California, Davis, May 1979 and May 1980*. Berkeley: Cooperative Extension, University of California, 1981.

Murdock, Catherine Gilbert. *Domesticating Drink: Women, Men, and Alcohol in America, 1870–1940*. Baltimore: Johns Hopkins University Press, 1998.

Muscadine, Doris, Maynard A. Amerine, and Bob Thompson, eds. *The University of California/Sotheby Book of California Wine*. Berkeley: University of California Press, 1984.

Nystrom, Justin A. *Creole Italian: Sicilian Immigrants and the Shaping of New Orleans Food Culture*. Athens: University of Georgia Press, 2018.

Okrent, Daniel. *Last Call: The Rise and Fall of Prohibition*. New York: Scribner, 2010.

Olmstead, Alan L. "Induced Innovation in American Agriculture." *Journal of Political Economy* 101 (February 1993): 100–118.

Olson, Nancy. *With a Lot of Help from Our Friends: The Politics of Alcoholism*. New York: Writers Club Press, 2003.

Ough, Cornelius S. *Winemaking Basics*. New York: Food Products Press, 1992.

Parsons, James J. "Uniqueness of California." *American Quarterly* 7 (Spring 1955).

Peele, Stanton. "The Conflict Between Public Health Goals and the Temperance Mentality." *American Journal of Public Health* 83 (June 1993): 805–10.

Peninou, Ernest P., and Gail G. Unzelman. *The California Wine Association and Its Member Wineries, 1894–1920*. Santa Rosa, CA: Nomis Press, 2000.

Peninou, Ernest P., and Sidney Greenleaf. *A Directory of California Wine Growers and Winemakers in 1860*. Berkeley: Tamalpais Press, 1967.

Perdue, Lewis. *The Wrath of Grapes: The Coming Wine Industry Shakeout and How to Take Advantage of It*. New York: Avon Books, 1999.

Phillips, Rod. *A Short History of Wine*. New York: HarperCollins, 2000.

———. *Wine: A Social and Cultural History of the Drink That Changed Our Lives*. London: Infinite Ideas, 2018.

Pickleman, Jack. "A Glass a Day Keeps the Doctor . . ." *American Surgeon* 56 (July 1990): 395–97.

Pinney, Thomas. *The City of Vines: A History of Wine in Los Angeles*. Berkeley: Heyday, 2017.

———. *A History of Wine in America: From Prohibition to the Present*. Berkeley: University of California Press, 2005.

———. *A History of Wine in America: From the Beginnings to Prohibition*. Berkeley: University of California Press, 1989.

———. *The Makers of American Wine: A Record of Two Hundred Years*. Berkeley: University of California Press, 2012.

Posert, Harvey, and Paul Franson. *Spinning the Bottle: Case Histories, Tactics and Stories of Wine Public Relations*. St. Helena, CA: HPPR Press, 2004.

Prial, Frank. *Decantations: Reflections on Wine by the New York Wine Critic Frank J. Prial*. New York: St. Martin's Press, 2001.

Renaud, E. B. "Wine, Alcohol, Platelets, and the French Paradox for Coronary Heart Disease." *Lancet* 339 (June 20, 1992): 1523–26.

Rich, Yale. *Cultural Exchange and the Cold War: Raising the Iron Curtain*. University Park: Pennsylvania State University Press, 2003.

Robertson, Carol. *The Little Red Book of Wine Law: A Case of Legal Issues*. Chicago: American Bar Association, 2008.

Rodeno, Michaela Kane. *Bubbles to Boardrooms: Serendipitous Stories from Inside the Wine Business*. Napa, CA: Villa Ragazzi Press, 2015.

Rolle, Andrew. *The Italian Americans: Troubled Roots*. New York: Free Press, 1980.

Rorabaugh, William J. *The Alcoholic Republic: An American Tradition*. New York: Oxford University Press, 1979.

———. *Prohibition: A Concise History*. New York: Oxford University Press, 2018.

Samsel, Lynn, Diane I. Hambley, and Raymond A. Marquardt. "Agribusiness' Competitiveness for Venture Capital." *Agribusiness* 7 (July 1991): 401–13.

Sbarboro, Andrea. *Temperance vs. Prohibition: Important Letters and Data from Our American Consuls, the Clergy and Other Eminent Men*. Leopold Classic Library. N.p., n.d.

Schnier, Robert F. "A Study of the Will to Survive of the Family-Owned Wineries of California." Master's thesis. Pepperdine University, 1982.

Schoonmaker, Frank. *Frank Schoonmaker's Encyclopedia of Wine*. New York: Hastings House, 1964.

Schoonmaker, Frank, and Tom Marvel. *The Complete Book of Wine*. New York: Duell, Sloan & Pearce, 1941.

Sensi-Isolani, Paola A., and Phylis Cancilla Martinelli, eds. *Struggle and Success: An Anthology of the Italian Immigrant Experience in California*. New York: Center for Immigration Studies, 1993.

Serlis, Harry G. *Wine in America*. New York: Newcomen Society in North America, 1972.

Shaw, David Scott. "Firm Export Strategies and Firm Export Performance in the United States Wine Industry: A Longitudinal Study." PhD diss., Purdue University, 1996.

Siler, Julia Flynn. *The House of Mondavi: The Rise and Fall of an American Wine Dynasty*. New York: Gotham Books, 2007.

Sims, Eric N. "A Study of the California Wine Industry and an Analysis of the Effects of the Canadian–United States Free Trade Agreement on the Wine Sector, with a Note on the Impact of the North American Free Trade Agreement on California Wine Exports." PhD diss., University of Arkansas, 1995.

Solana Rosillo, Juan B. "Firm Strategies in International Markets: The Case of International Entry into the United States Wine Industry (Market Entry, International Trade)." PhD diss., Purdue University, 1997.

Sommers, Brian J. *The Geography of Wine: How Landscapes, Cultures, Terroir, and the Weather Make a Good Drop*. New York: Plume, 2008.

Sosnowski, Vivienne. *When the Rivers Ran Red: An Amazing Story of Courage and Triumph in America's Wine Country*. New York: Palgrave Macmillan, 2009.

Spitze, R.G. F. "A Continuing Evolution in U.S. Agricultural Policy." *Agricultural Economics* 77 (1990): 126–39.

Stuller, Jay, and Glen Martin. *Through the Grapevine: The Real Story Behind America's $8 Billion Wine Industry*. New York: HarperCollinsWest, 1994.

Sullivan, Charles L. *A Companion to California Wine: An Encyclopedia of Wine and Winemaking from the Mission Period to the Present*. Berkeley: University of California Press, 1998.

———. *Like Modern Edens: Winegrowing in Santa Clara Valley and Santa Cruz Mountains, 1798–1891*. Cupertino: California History Center, 1982.

———. *Napa Wine: A History from Mission Days to Present*. San Francisco: Wine Appreciation Guild, 1994.

Taber, George M. *Judgement of Paris: California vs. France and the Historic 1976 Paris Tasting That Revolutionized Wine*. New York: Scribner, 2005.

———. *To Cork or Not to Cork: Tradition, Romance, Science, and the Battle for the Wine Bottle*. New York: Scribner, 2007.

Teiser, Ruth, and Catherine Harroun. *Winemaking in California*. New York: McGraw-Hill, 1983.

Trubek, Amy B. *The Taste of Place: A Cultural Journey into Terroir*. Berkeley: University of California Press, 2008.

Tyrrell, Ian R. *Sobering Up: From Temperance to Prohibition in Antebellum America, 1800–1860*. Westport, CT: Greenwood Press, 1979.

Unwin, Tim. *Wine and the Vine: An Historical Geography of Viticulture and the Wine Trade*. New York: Routledge, 1991.

U.S. News and World Report: The Wine Marketing Handbook, 1972. New York: Gavin-Jobson, 1972.

Wait, Frona Eunice. *Wines and Vines of California; or, A Treatise on the Ethics of Wine Drinking*. 1889. Reprint. Berkeley: Howell-North Books, 1973.

Warner, Nicholas O. *Spirits of America: Intoxication in Nineteenth-Century American Literature*. Norman: University of Oklahoma Press, 1997.

Waters, Alejandro. "Rebuilding Technologically Competitive Industries: Lessons from Chile's and Argentina's Wine Industry Restructuring." PhD diss., Massachusetts Institute of Technology, 1999.

Welsh, S., C. Davis, and A. Shaw. "A Brief History of Food Guides in the United States." *Nutrition Today* (November–December 1992): 6–11.

———. "Development of the Food Guide Pyramid." *Nutrition Today* (November–December 1992).

White, Anthony Gene. "State Policy and Public Administration Impacts on an Emerging Industry: The Wine Industry in Oregon and Washington." PhD diss., Portland State University, 1993.

Yandle, Bruce. "Bootleggers and Baptists: The Education of a Regulatory Economist." *Regulation* 7, no. 3 (1983): 12–16.

Yandle, Bruce, and Adam Smith. *Bootleggers and Baptists: How Economic Forces and Moral Persuasion Interact to Shape Regulatory Politics*. Washington, DC: CATO Institute, 2014.

Index

About the Author

Victor W. Geraci, who earned his PhD in history at the University of California, Santa Barbara, was an assistant and then associate professor of history at Central Connecticut State University from 1997 to 2003. In 2003, Geraci began a new role with the University of California, Berkeley, Bancroft Library Oral History Center as a food and wine historian/specialist, and he was the program's associate director until his retirement in 2013. His main area of research includes American agriculture with a specific focus on the California wine industry.